walkermaths 2.2

GRAPHS

NCEA Level 2 Internal

Charlotte Walker and Victoria Walker

Walker Maths 2.2 Graphs
1st Edition
Charlotte Walker
Victoria Walker

Designer: Cheryl Smith, Macarn Design
Production controller: Siew Han Ong

Any URLs contained in this publication were checked for currency during the production process. Note, however, that the publisher cannot vouch for the ongoing currency of URLs.

Acknowledgements
Cover photo courtesy of Shutterstock.

We wish to thank the Boards of Trustees of Darfield and Riccarton High Schools for allowing us to use materials and ideas developed while teaching. Our thanks also go to all past and present colleagues who have generously shardd their expertise and ideas.

For product information and technology assistance,
in Australia call **1300 790 853**;
in New Zealand call **0800 449 725**

For permission to use material from this text or product, please email
aust.permissions@cengage.com

National Library of New Zealand Cataloguing-in-Publication Data
A catalogue record for this book is available from the National Library of New Zealand.

978 0 17 041598 9

Cengage Learning Australia
Level 7, 80 Dorcas Street
South Melbourne, Victoria Australia 3205

Cengage Learning New Zealand
Unit 4B Rosedale Office Park
331 Rosedale Road, Albany, North Shore 0632, NZ

For learning solutions, visit **cengage.co.nz**

Printed in China by 1010 Printing International Limited.
7 8 24

CONTENTS

 Glossary

Make your own glossary of key terms:

Term	Definition	Picture/Example
Function		
Properties of functions		
Features of functions		
Intercept		
Asymptote		
Local maximum or minimum		
Periodic function		
Cycle		
Amplitude		
Period		

 ISBN: 9780170415989

Term	Definition	Picture/Example
Logarithmic		
Cubic		
Quartic		
Rectangular hyperbola		
Absolute value		
Exponential		
Domain		
Range		
Real numbers		
Converse of a function		

Functions

- A function can be thought of as a box that does **exactly the same thing** to every value fed into it.
- For every value fed into the box, there can be no more than **one** output value.

Input x		Output f(x)
1	→ 🔲 →	12
2	→ 🔲 →	18
3	→ 🔲 →	28
4	→ 🔲 →	42
–4	→ 🔲 →	42
x	→ 🔲 →	$2x^2 + 10$

Notice that there is **one** output value for every input value.

However, different input values may give the same output value.

In this case, every input value was
- squared,
- the squared value was doubled,
- and then 10 was added.

- The same function can also be expressed on a graph.

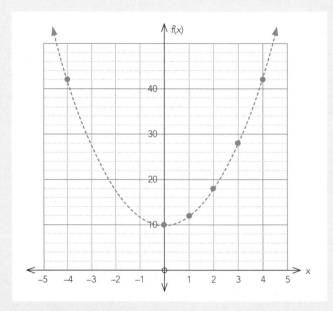

- Until now, you have used expressions such as $y = 2x^2 + 10$.
  It is more useful to use the expression $f(x) = 2x^2 + 10$.
- This can be used to write expressions such as $f(5) = 2(5)^2 + 10 = 60$

 ISBN: 9780170415989

The vertical line test

- Because every input value can produce no more than one output value, an easy test to find out whether a relationship is a function is to draw a **vertical line**.

Cuts in **one** place $\Rightarrow$ **is** a function.

Cuts in **more than** one place $\Rightarrow$ **is not** a function.

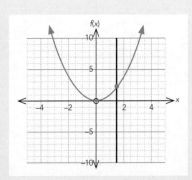

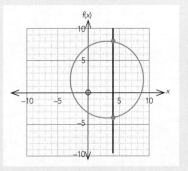

Domain and range

Domain = set of all input values (x values).
Range = set of all output values ($f(x)$ or y).

Examples:

$$f(x) = x^2 + 2$$

$$(x - 3)^2 + (y - 2)^2 = 36$$

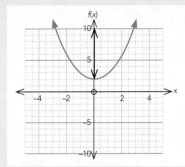

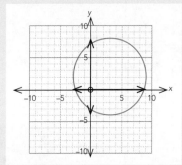

Domain: The real numbers
Range: $f(x) \geq 2$

Domain: $-3 \leq x \leq 9$
Range: $-4 \leq y \leq 8$

The rectangular hyperbolae

- You will meet these graphs later in the book.
- They have **asymptotes** (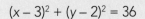).
- An asymptote is a line that the graph gets very close to, but **never reaches**.

Example: $(x - 3)(y + 2) = 6$

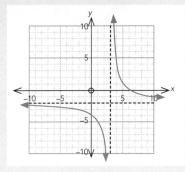

Domain: The real numbers **except for 3**.
Range: The real numbers **except for –2**.

For each of the following graphs, state whether or not they are functions, and if they are, give the domain and range.

1

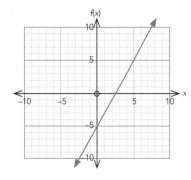

$f(x) = 2x - 5$

Function?   Yes/No

Domain: _____

Range: _____

2

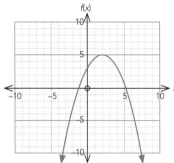

$f(x) = -\frac{1}{2}(x - 2)^2 + 5$

Function?   Yes/No

Domain: _____

Range: _____

3

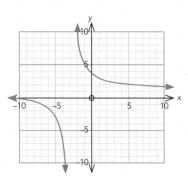

$(x + 3)(y - 1) = 8$

Function?   Yes/No

Domain: _____

Range: _____

4

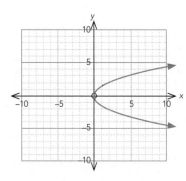

$x = \frac{1}{2}y^2$

Function?   Yes/No

Domain: _____

Range: _____

5

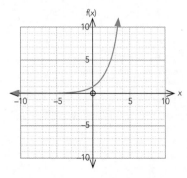

$f(x) = 2^x$

Function?   Yes/No

Domain: _____

Range: _____

Calculations using function notation

Examples:

1 Evaluate the following when $f(x) = x^2 - 3x + 7$:

$f(2) = (2)^2 - 3(2) + 7 = 5$

$f(0) = (0)^2 - 3(0) + 7 = 7$

$f(-4) = (-4)^2 - 3(-4) + 7 = 35$

Wherever there is an x, replace it with a 2.

2 Evaluate the following when $f(x) = \dfrac{6}{x}$:

$f(3) = \dfrac{6}{3} = 2$

$f(0) = \dfrac{6}{0}$ ← This is undefined.

3 For what value(s) of x will $f(x) = 0$ if $f(x) = x^2 + 2x - 63$?

$f(x) = (x - 7)(x + 9) = 0$

$\therefore x = 7$ or -9

Calculate the values of the following functions.

1 $f(x) = 3x - 4$

$f(2) =$ _____

$f(-4) =$ _____

$f(0) =$ _____

2 $f(x) = x^2 + 4x - 3$

$f(2) =$ _____

$f(-4) =$ _____

$f(0) =$ _____

3 $f(x) = 2^x$

$f(2) =$ _____

$f(-4) =$ _____

$f(0) =$ _____

Solve the following.

4 For what value(s) of x will $f(x) = 0$ when $f(x) = 8x - 100$?

_____

_____

_____

5 For what value(s) of x will $f(x) = 0$ when $f(x) = x^2 - 3x - 10$?

_____

_____

_____

Features of functions

Intercepts

- These are the points where a line or curve **crosses the x and y axes**.
- Sometimes there are no intercepts, or there may be several.

Example: $f(x) = (x - 3)(x - 2)(x + 1)$

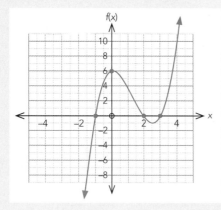

x intercepts: (–1, 0), (2, 0) and (3, 0)

y intercept: (0, 6)

Maximum and minimum points

- These are the **highest** and **lowest** points, respectively, reached by a graph.
- Where graphs extend up or down indefinitely, these points do not exist.
- A **local** maximum or minimum is the highest or lowest point, respectively, reached by a graph at points where it changes vertical direction, but they are not the highest or lowest point on the entire graph.

Examples:

$y = \sin(x)$

$f(x) = x^3 + 1.5x^2 - 6x = 0$

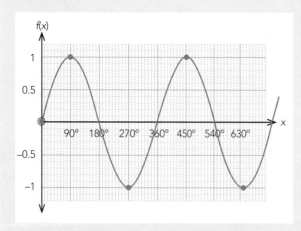

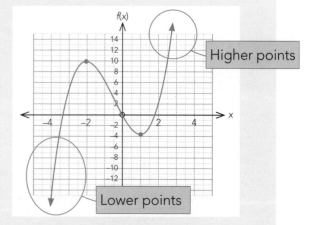

Higher points

Lower points

Maximum: (90°, 1), (450°, 1), etc

Minimum: (270°, –1), (630°, –1), etc

Local maximum: (–2, 10)

Local minimum: (1, –3.5)

Maximum: no maximum

Minimum: no minimum

   ISBN: 9780170415989

Asymptotes

- An asymptote is a line that the graph approaches, but **never reaches**.
- You will deal with horizontal and vertical asymptotes.

Example: $(y - 2)(x + 3) = 6$

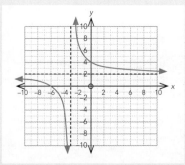

Vertical asymptote: $x = -3$

Horizontal asymptote: $y = 2$

Symmetry

Reflective symmetry
- A function has reflective symmetry if it can be reflected in a horizontal or vertical line, and remain unchanged.
- Alternatively, one function may be a reflection of another function.

Examples:

$xy = 12$

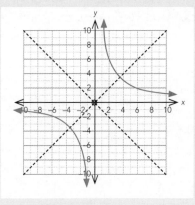

The two parts of this function show **reflective symmetry** in the line $y = -x$.

Each part also shows **reflective symmetry** in the line $y = x$.

$f(x) = x^2 + 5$ and $f(x) = -x^2 + 1$

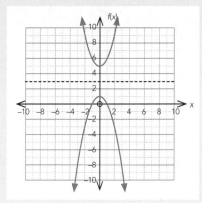

$f(x) = -x^2 + 1$ is a **reflection** of $f(x) = x^2 + 5$ in the line $f(x) = 3$.

Rotational symmetry

- Some functions exhibit point or rotational symmetry.

Examples: $f(x) = (x + 1)(x - 1)(x + 3) + 3$

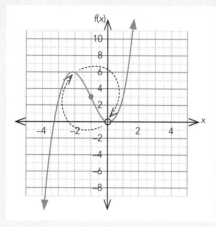

This graph shows **rotational symmetry** of order 2 about the **point** (–1, 3).

$(y - 2)(x + 3) = 6$

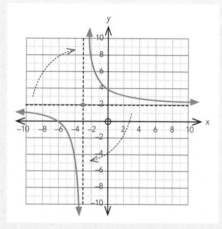

All rectangular hyperbolas have **rotational symmetry** of order 2 about the point where the asymptotes meet.

$f(x) = \sin(x)$

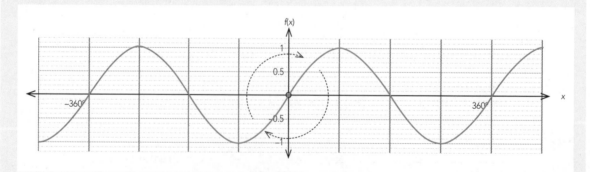

All trigonometric graphs show **rotational symmetry** of order 2.

    ISBN: 9780170415989

Features associated with periodic functions

Periodic functions
- $f(x) = \sin(x)$, $f(x) = \cos(x)$ and $f(x) = \tan(x)$ are all **periodic functions**.
- Periodic functions **repeat the same pattern indefinitely** to the left and to the right of the $f(x)$ axis.

Example: $f(x) = \sin(x)$

Cycle = **one complete pattern**, with no repetition.

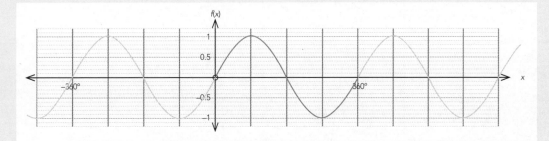

Period = horizontal **distance** required for one complete cycle.
The periods for $f(x) = \sin(x)$, $f(x) = \cos(x)$ $f(x) = \tan(x)$ are all 360°.

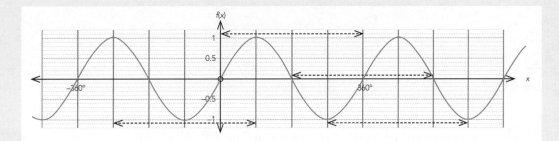

Amplitude = **height** from the centre line to the peak or trough.

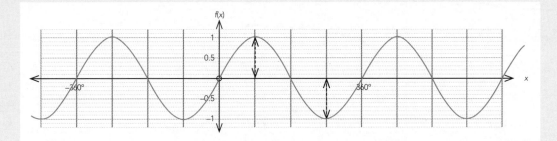

The amplitudes for $f(x) = \sin(x)$ and $f(x) = \cos(x)$ are both 1.

Write down the features of the following graphs.

1 $f(x) = -x^3 + 3x + 2$

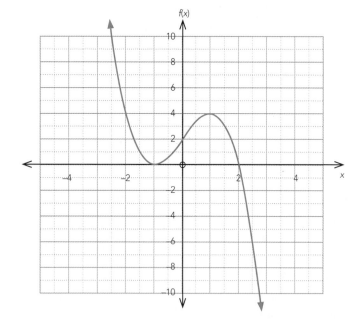

Maximum: _____

Minimum: _____

x intercept: _____

y intercept: _____

Asymptotes: _____

Axes of symmetry: _____

_____

Centre of rotational symmetry:

_____

Domain: _____

Range: _____

2 $x(y - 1) = -6$

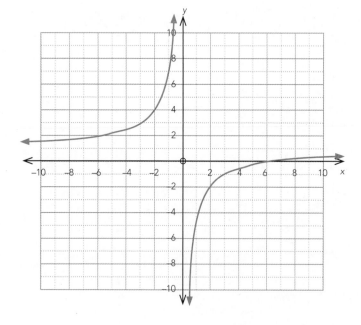

Maximum: _____

Minimum: _____

x intercept: _____

y intercept: _____

Asymptotes: _____

Axes of symmetry: _____

_____

Centre of rotational symmetry:

_____

Domain: _____

Range: _____

ISBN: 9780170415989

3 $y = 3^x + 2$

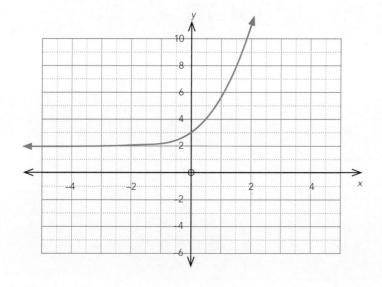

Maximum: _____

Minimum: _____

x intercept: _____

y intercept: _____

Asymptotes: _____

Axes of symmetry: _____

_____

Centre of rotational symmetry:

_____

Domain: _____

Range: _____

4 $f(x) = 3\sin(2x)$

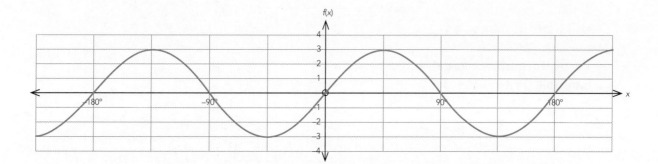

Maximum: _____

Minimum: _____

x intercept: _____

y intercept: _____

Asymptotes: _____

Axes of symmetry: _____

_____

Centre of rotational symmetry:

_____

Domain: _____

Range: _____

Amplitude: _____

Period: _____

Parabolas: revision

- For this standard, you will need to be completely familiar with quadratic equations and drawing graphs of parabolas.
- A summary is given below.

The basic parabola

$f(x) = x^2$

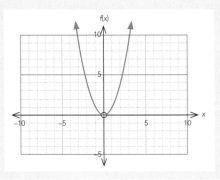

Transformations of parabolas

Translations of parabolas

1 Vertical translations: $f(x) = x^2 \pm c$

Example 1: $f(x) = x^2 + 3$ **Example 2:** $f(x) = x^2 - 6$

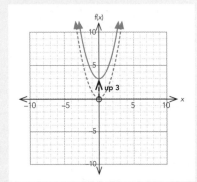
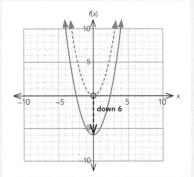

2 Horizontal translations: $f(x) = (x \pm b)^2$

Example 1: $f(x) = (x + 4)^2$ **Example 2:** $f(x) = (x - 5)^2$

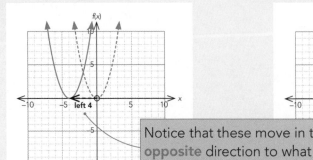
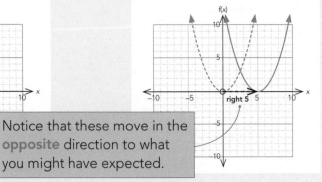

Notice that these move in the **opposite** direction to what you might have expected.

    ISBN: 9780170415989

3 Combinations: $f(x) = (x \pm b)^2 \pm c$

Example 1: $f(x) = (x - 3)^2 + 4$

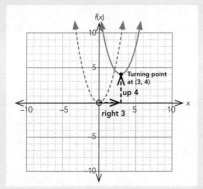

Example 2: $f(x) = (x + 5)^2 - 6$

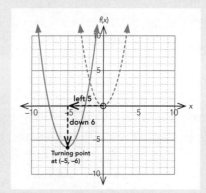

Reflections of parabolas

1 Vertical reflections: $f(x) = -(x \pm b)^2 \pm c$

Example 1: $f(x) = -x^2$

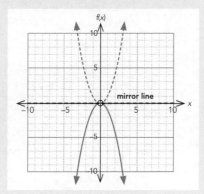

Example 2: $f(x) = -(x + 5)^2$

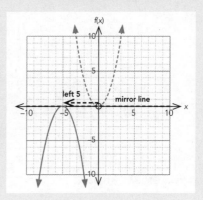

2 Horizontal reflections:

All parabolas reflect onto themselves when reflected horizontally through their axis of symmetry.

Enlargements of parabolas

Vertical enlargements: $f(x) = ax^2$

$a > 1 \Rightarrow$ function is **stretched** vertically

Example 1: $f(x) = 2x^2$

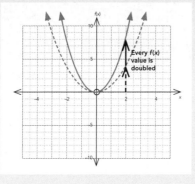

$a < 1 \Rightarrow$ function is **shortened** vertically

Example 2: $f(x) = \dfrac{1}{2}x^2$

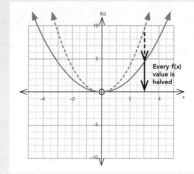

Plotting parabolas with the form $y = a(x \pm p)(x \pm q)$

Examples:

1 Draw the graph of $y = (x + 1)(x - 3)$.

Method 1: Do a table. This will *always* work.

Method 2: If $y = (x + p)(x - q)$, plot:
1. x intercepts at -p and +q.
2. y intercept at (+p) x (-q) or where $x = 0$.
3. Axis of symmetry halfway between the x intercepts.
4. Substitute x coordinate of axis of symmetry into equation to find turning point.
5. Substitute to find extra points.
6. Use the axis of symmetry and reflection to plot missing points.

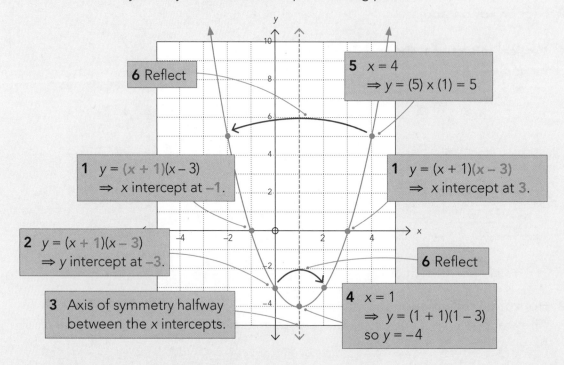

6 Reflect

5 $x = 4$
$\Rightarrow y = (5) \times (1) = 5$

1 $y = (x + 1)(x - 3)$
$\Rightarrow x$ intercept at -1.

1 $y = (x + 1)(x - 3)$
$\Rightarrow x$ intercept at 3.

2 $y = (x + 1)(x - 3)$
$\Rightarrow y$ intercept at -3.

6 Reflect

3 Axis of symmetry halfway between the x intercepts.

4 $x = 1$
$\Rightarrow y = (1 + 1)(1 - 3)$
so $y = -4$

2 Draw the graph of $y = \frac{1}{2}x(5 - x)$.

1. x intercepts at 0 and 5.

2. y intercept at $\frac{1}{2}(0) \times (5) = 0$.

3. Axis of symmetry at $x = 2.5$.

4. Turning point at
$y = \frac{1}{2} \times 2\frac{1}{2} \times 2\frac{1}{2} = 3\frac{1}{8}$

5. Extra point:
$x = 1 \rightarrow y = \frac{1}{2} \times 1 \times 4 = 2$

$x = 2 \rightarrow y = \frac{1}{2} \times 2 \times 3 = 3$

$x = 6 \rightarrow y = \frac{1}{2} \times 6 \times -1 = -3$

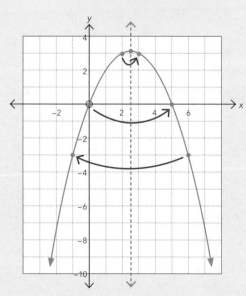

6. See arrows.

 ISBN: 9780170415989

Draw graphs for the following equations.

1 $y = -2(x - 3)(x + 1)$

    1 x intercepts: _____

    2 y intercept: _____

    3 Axis of symmetry at x = _____

    4 Turning point at

       y = _____

    5 Extra points:

       x = _____ → y =

       x = _____ → y =

    6 Reflections

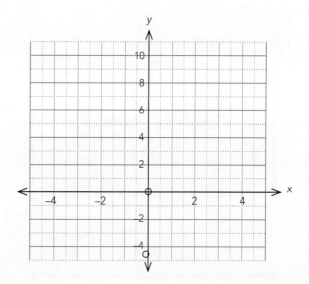

2 $y = \frac{1}{2}x(x - 4)$

    1 x intercepts: _____

    2 y intercept: _____

    3 Axis of symmetry at x = _____

    4 Turning point at

       y = _____

    5 Extra points:

       x = _____ → y =

       x = _____ → y =

    6 Reflections

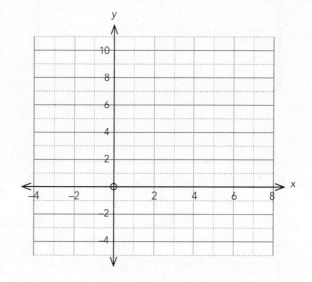

3 $y = -\frac{1}{4}(x + 3)(x - 4)$

    1 x intercepts: _____

    2 y intercept: _____

    3 Axis of symmetry at x = _____

    4 Turning point at

       y = _____

    5 Extra points:

       x = _____ → y =

       x = _____ → y =

    6 Reflections

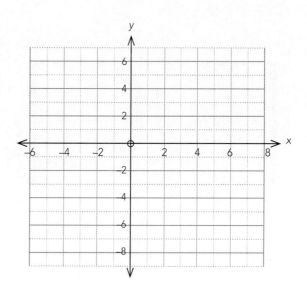

Using a graphics calculator to draw graphs

Draw the graph of $y = \frac{1}{2}x(5 - x)$.

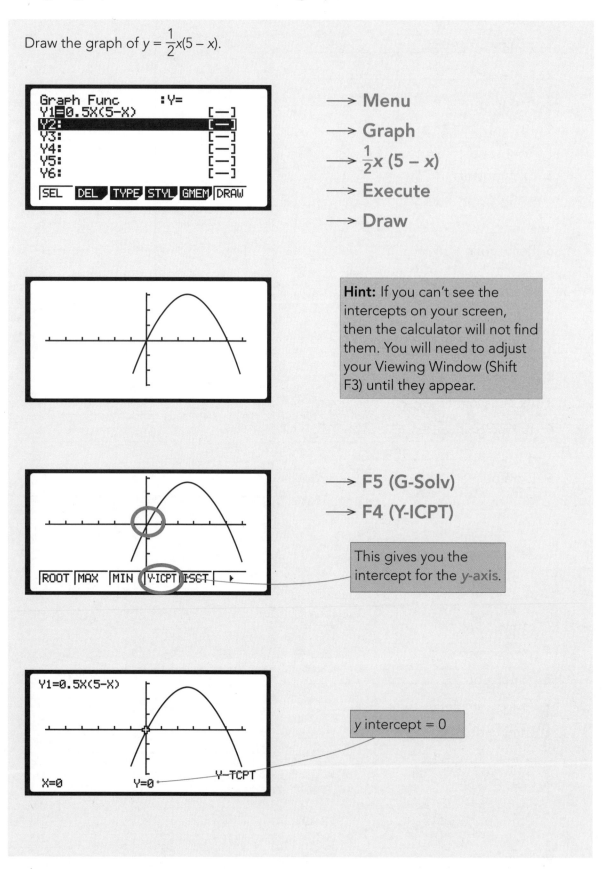

→ Menu

→ Graph

→ $\frac{1}{2}x(5 - x)$

→ Execute

→ Draw

Hint: If you can't see the intercepts on your screen, then the calculator will not find them. You will need to adjust your Viewing Window (Shift F3) until they appear.

→ F5 (G-Solv)

→ F4 (Y-ICPT)

This gives you the intercept for the **y-axis**.

y intercept = 0

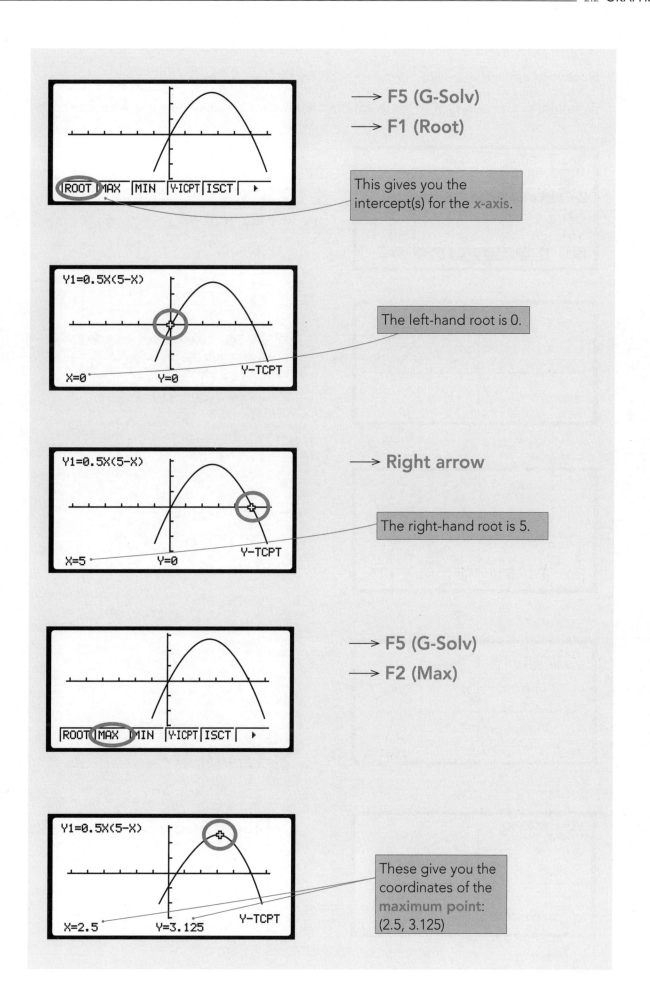

→ **F5 (G-Solv)**

→ **F1 (Root)**

This gives you the intercept(s) for the x-axis.

The left-hand root is 0.

→ **Right arrow**

The right-hand root is 5.

→ **F5 (G-Solv)**

→ **F2 (Max)**

These give you the coordinates of the maximum point: (2.5, 3.125)

Note: You can add more lines or curves, and find where they meet.

Example: Add $y = -\frac{1}{2}x + 4$ to the graph, and find where it meets the curve $y = \frac{1}{2}x(5 - x)$.

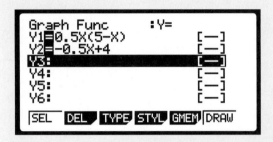

$\longrightarrow$ Exit

$\longrightarrow -\frac{1}{2}x + 4$

$\longrightarrow$ Execute

$\longrightarrow$ Draw

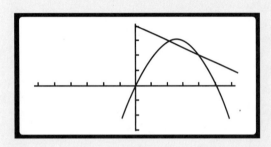

Hint: Once again, you may need to adjust your Viewing Window (Shift F3) until you can see everything you need.

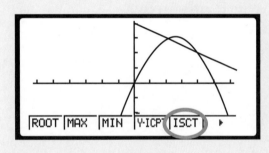

$\longrightarrow$ F5 (G-Solv)

$\longrightarrow$ F5 (ISCT)

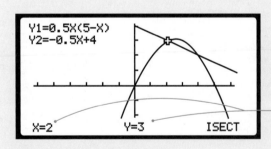

The coordinates of the left point of intersection are (2, 3).

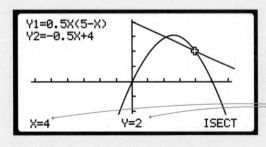

$\longrightarrow$ Right arrow

The coordinates of the right point of intersection are (4, 2).

   ISBN: 978017041598

Writing equations for parabolas

- Start by marking integral points that you are **certain** the parabola passes through. (Do not assume a graph passes through a point just because it **looks** as though it does.)
- It will pass through

either the x intercepts 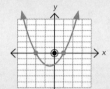 **or** the turning point.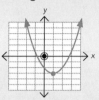

1 The x intercepts

The equation takes the form $y = a(x \pm p)(x \pm q)$.

Example:
1

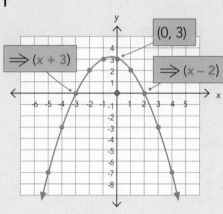

Steps:
1 Mark all the integral points ($\bullet$).
2 x intercepts $\Rightarrow$ equation must be
$$y = a(x - 2)(x + 3)$$
3 Substitute the coordinates of another integral point, e.g. (0, 3):
$$3 = a(0 - 2)(0 + 3)$$
$$3 = -6a$$
$$\therefore a = -\frac{1}{2}$$
So the equation must be $y = -\frac{1}{2}(x - 2)(x + 3)$.

2 The turning point

The equation takes the form $y = a(x \pm b)^2 \pm c$.

Example:
1

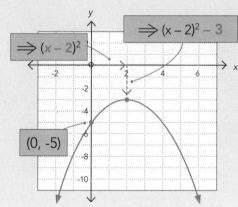

Steps:
1 Mark the turning point and y intercept ($\bullet$).
2 Turning point moved **2 units to right**
$$\Rightarrow y = a(x - 2)^2 \pm c$$
3 Turning point moved **3 units down**
$$\Rightarrow y = a(x - 2)^2 - 3$$
4 Substitute the coordinates of the y intercept, (0, −5):
$$-5 = a(0 - 2)^2 - 3$$
$$-2 = 4a$$
$$a = -\frac{1}{2}$$
So the equation must be $y = -\frac{1}{2}(x - 2)^2 - 3$.

Write equations for the following parabolas. Integral points are marked with a dot.

4

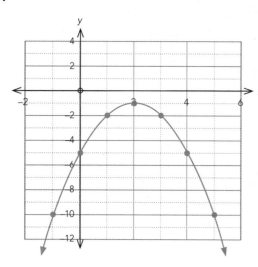

a Turning point moved ____ unit(s) to right/left

$\rightarrow\ y = a(x\_\_\_\_\_)^2 \pm c$

b Turning point moved ____ unit(s) up/down

$\rightarrow\ y = a(x\_\_\_\_\_)^2$ _____

c To find a: substitute the coordinates of the y intercept, (0, ____):

_____

_____

So the equation must be

$y =$ _____

5

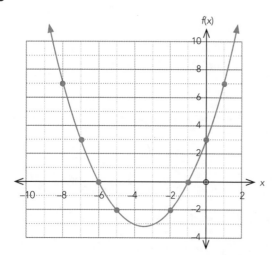

a x intercepts $\rightarrow$ equation must be

$f(x) = a(_____)(_____)$

b To find a: substitute the coordinates of another integral point:

_____

_____

So the equation must be

$f(x) =$ _____

6

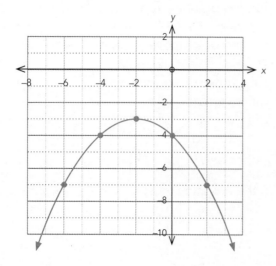

_____

_____

_____

_____

_____

_____

_____

_____

_____

7

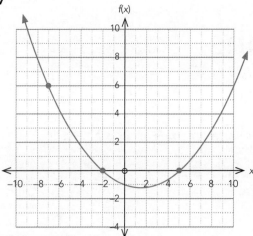

8

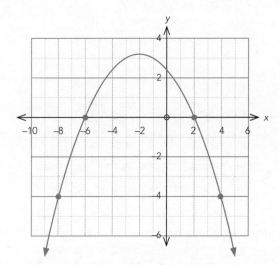

9

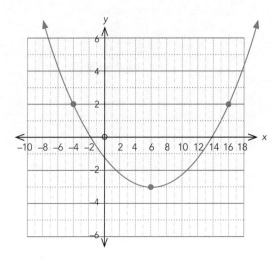

Plotting graphs

- If you are given the equation of a graph, you can **always** create a **table** and use that to plot some points through which it passes.
- You will have used this method to find graphs for straight lines and parabolas in previous courses.

Other graphs that you will meet in this book are:

1 Cubic graphs

- These are graphs whose equation includes an **x^3**.

Example 1: $f(x) = x^3$

x	f(x) = x³
3	27
2	8
1	1
0	0
−1	−1
−2	−8
−3	−27

It is important to find where the graph cuts the axes.

Plotting points for x values of −3 to 3 shows what this graph looks like. However, often you will need to think carefully about which values you plot.

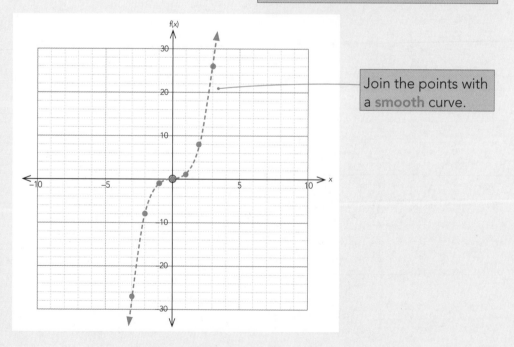

Join the points with a **smooth** curve.

Example 2: $f(x) = x(x - 2)(x + 3)$

x	$f(x) = x(x - 2)(x + 3)$
3	18
2	0
1	−4
0	0
−1	6
−2	8
−3	0
−4	−24

Where $x = 2$,
$f(x) = x(x - 2)(x + 3) = 0$

Where $x = 0$,
$f(x) = x(x - 2)(x + 3) = 0$

Where $x = -3$,
$f(x) = x(x - 2)(x + 3) = 0$

It has been necessary to plot the point where $x = -4$ in order to complete the picture.

Notice that the turning points are not always halfway between the intercepts.

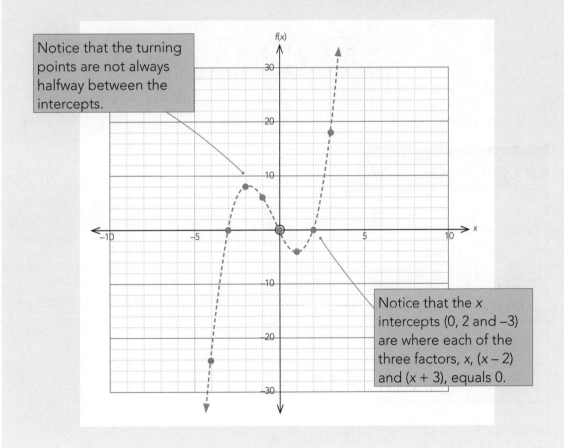

Notice that the x intercepts (0, 2 and −3) are where each of the three factors, x, (x − 2) and (x + 3), equals 0.

Complete the following tables and graphs.

1   **a** $f(x) = x^2(x - 4)$

x	$f(x) = x^2(x - 4)$
5	
4	
3	–9
2	–8
1	–3
0	0
–1	
–2	
–3	(–63)

Needed to complete the graph.

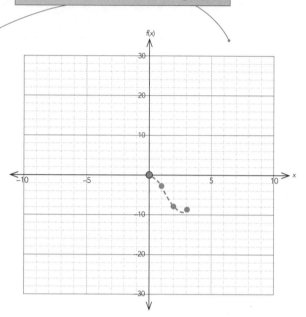

Some values for $f(x)$ will be too big to plot, but may be useful to show where the graph is heading.

b $f(x) = (x + 1)(x - 1)(x - 3)$

x	$f(x) = (x + 1)(x - 1)$ $(x - 3)$
3	
2	
1	
0	3
–1	0
–2	–15
–3	

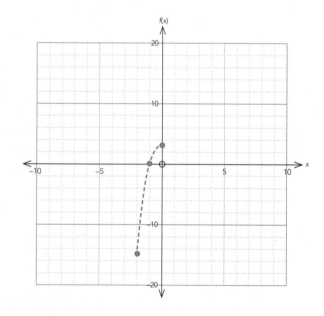

2 Quartic graphs

- These are graphs whose equation includes an x^4.

Example 1: $f(x) = x^4$

x	$f(x) = x^4$
3	81
2	16
1	1
0	0
−1	1
−2	16
−3	81

You will find that values for $f(x)$ become too large to plot.

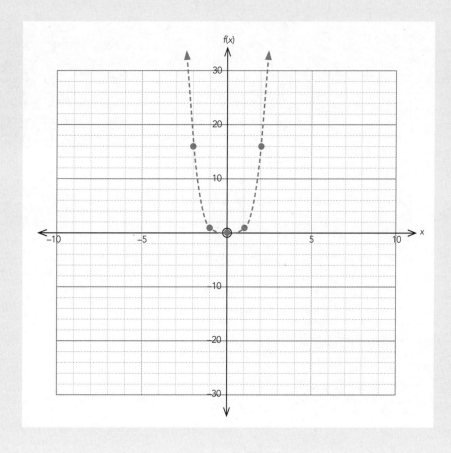

Example 2: $f(x) = x(x - 1)(x + 2)(x - 4)$

x	$f(x) = x(x - 1)(x + 2)(x - 4)$
4	0
3	–30
2	–16
1	0
0	0
–1	–10
–2	0
–3	(84)

Where $x = 4$,
$f(x) = x(x - 1)(x + 2)(x - 4) = 0$

Where $x = 1$,
$f(x) = x(x - 1)(x + 2)(x - 4) = 0$

Where $x = 0$,
$f(x) = x(x - 1)(x + 2)(x - 4) = 0$

Where $x = -2$,
$f(x) = x(x - 1)(x + 2)(x - 4) = 0$

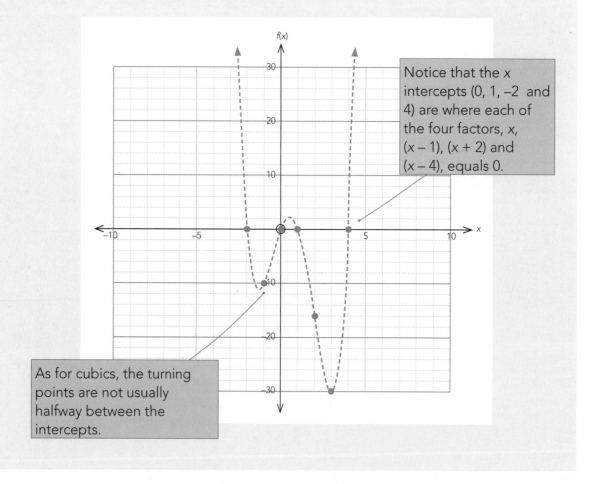

Notice that the x intercepts (0, 1, –2 and 4) are where each of the four factors, x, $(x - 1)$, $(x + 2)$ and $(x - 4)$, equals 0.

As for cubics, the turning points are not usually halfway between the intercepts.

 ISBN: 9780170415989

Complete the following tables and graphs.

2 **a** $f(x) = x^2(x - 2)(x + 3)$

x	$f(x) =$ $x^2(x - 2)(x + 3)$
3	
2	
1	−4
0	0
−1	−6

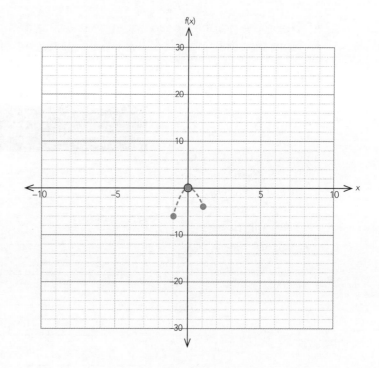

b $f(x) = (x + 1)(x + 2)(x - 1)(x - 3)$

x	$f(x) = (x + 1)(x + 2)$ $(x - 1)(x - 3)$

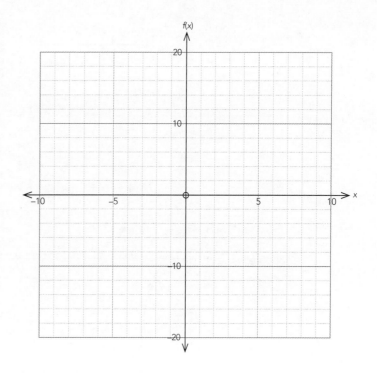

3 Rectangular hyperbolae

- These are graphs whose equation is of the form $f(x) = \dfrac{constant}{x}$ or $xy = $ **constant**.
- Note that the singular of hyperbolae is hyperbola.

Example: $f(x) = \dfrac{2}{x}$ or $y = \dfrac{2}{x}$ or $xy = 2$

x	$f(x) = \dfrac{2}{x}$
8	$\dfrac{1}{4}$
4	$\dfrac{1}{2}$
2	1
1	2
0	Undefined
−1	−2
−2	−1
−4	$-\dfrac{1}{2}$
−8	$-\dfrac{1}{4}$

Choose your x values carefully, otherwise you might get messy values for f(x).

The product of x and f(x) must always be 2.

Where f(x) is undefined, there is a vertical asymptote.

The line f(x) = x forms an axis of symmetry (------), so you do not need to calculate the coordinates of **every** point: you can reflect some in this line.

The x and f(x) axes form asymptotes.

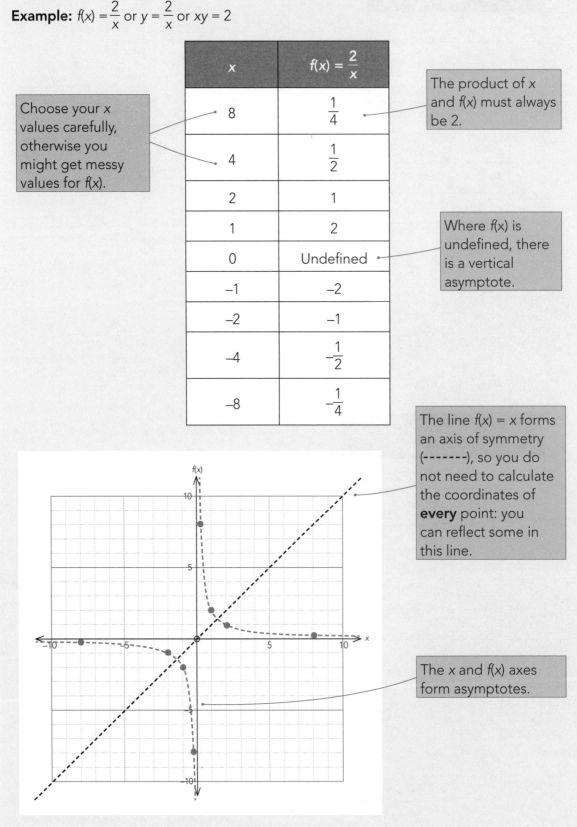

   ISBN: 9780170415989

Complete the following tables and graphs.

3   **a** $f(x) = \dfrac{6}{x}$

x	$f(x) = \dfrac{6}{x}$
3	
2	
1	6
0	Undefined
−1	−6
−2	
−3	

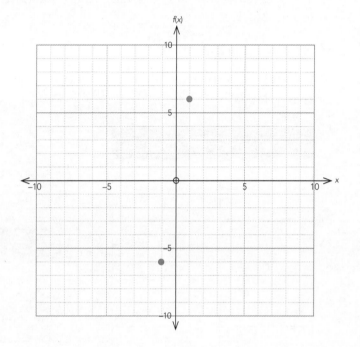

b $f(x) = \dfrac{10}{x}$

x	$f(x) = \dfrac{10}{x}$

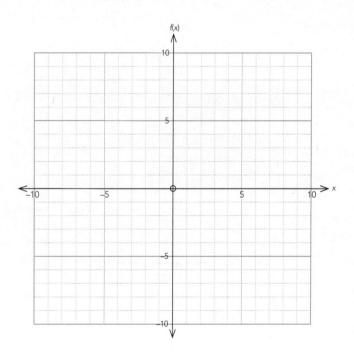

4 Exponential graphs

- These take the form $f(x) = $ (base)^x where the base can be any **positive** number (but not 0).

These can be of two types:

a Where **base > 1**; these are called **growth curves**.
   Example: $f(x) = 2^x$

x	$f(x) = 2^x$
3	8
2	4
1	2
0	1
−1	$\frac{1}{2}$
−2	$\frac{1}{4}$
−3	$\frac{1}{8}$

If x increases by 1, then $f(x)$ **doubles**.

(Any number)^x will pass through (0, 1) because (anything)^0 = 1.

Remember that $p^{-1} = \frac{1}{p^1}$, so $2^{-1} = \frac{1}{2}$

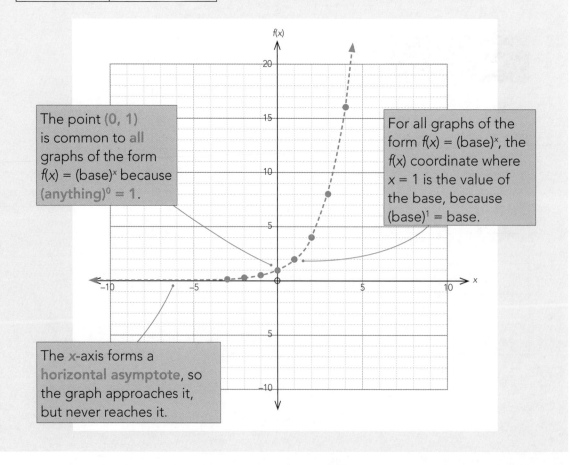

The point (0, 1) is common to **all** graphs of the form $f(x) = $ (base)^x because **(anything)^0 = 1**.

For all graphs of the form $f(x) = $ (base)^x, the $f(x)$ coordinate where $x = 1$ is the value of the base, because (base)^1 = base.

The x-axis forms a **horizontal asymptote**, so the graph approaches it, but never reaches it.

Complete the following tables and graphs.

4   **a**   **i** $f(x) = 3^x$

x	$f(x) = 3^x$
1	3
0	1
−1	$\frac{1}{3}$

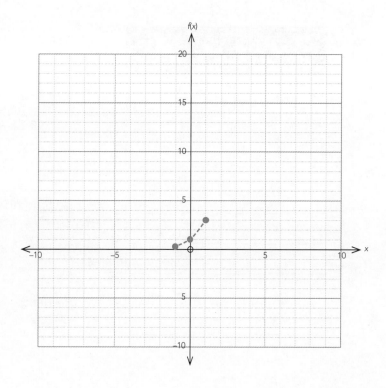

ii $f(x) = 10^x$

x	$f(x) = 10^x$
3	
2	
1	10
0	1
−1	0.1 or $\frac{1}{10}$
−2	
−3	

You will not be able to plot these points because of the scale.

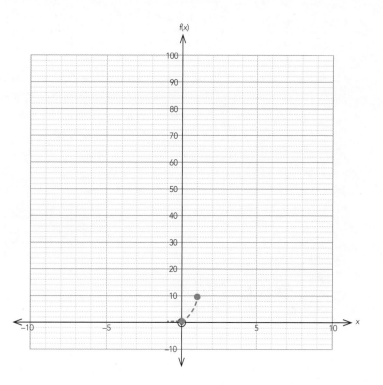

b Where **0 < base < 1**; these are called **decay curves**.

Example: $f(x) = \left(\frac{1}{2}\right)^x$

x	$f(x) = \left(\frac{1}{2}\right)^x$
3	$\frac{1}{8}$
2	$\frac{1}{4}$
1	$\frac{1}{2}$
0	1
−1	2
−2	4
−3	8

If x increases by 1, then f(x) **halves**.

(Any number)^x will pass through (0, 1) because (anything)^0 = 1.

Remember that $\left(\frac{1}{2}\right)^{-1} = \left(\frac{2}{1}\right)^1 = 2$

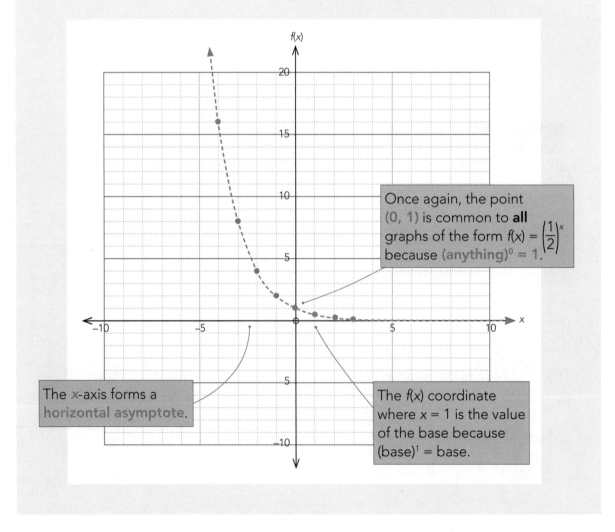

Once again, the point **(0, 1)** is common to **all** graphs of the form $f(x) = \left(\frac{1}{2}\right)^x$ because **(anything)^0 = 1**.

The **x**-axis forms a **horizontal asymptote**.

The f(x) coordinate where x = 1 is the value of the base because (base)^1 = base.

   ISBN: 9780170415989

Complete the following tables and graphs.

4   **b i** $f(x) = \left(\dfrac{1}{4}\right)^{x}$

x	$f(x) = \left(\dfrac{1}{4}\right)^{x}$
3	
2	
1	$\dfrac{1}{4}$
0	1
–1	4
–2	
–3	

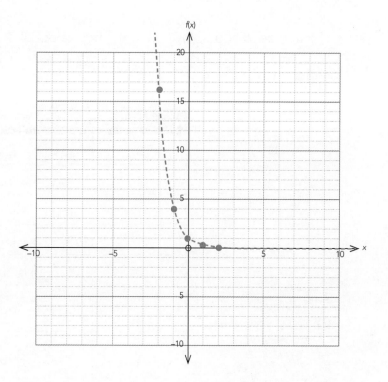

ii $f(x) = \left(\dfrac{1}{10}\right)^{x}$

x	$f(x) = \left(\dfrac{1}{10}\right)^{x}$
3	
2	
1	
0	1
–1	10
–2	
–3	

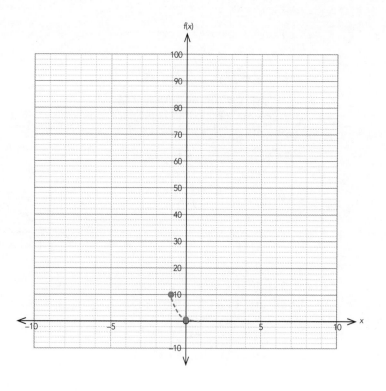

5 Logarithmic graphs

- Remember that $y = \log_{base}x$ is the equivalent of $x = base^y$.

Example: $y = \log_2 x$ or $x = 2^y$

For $x = 8$:
$$y = \log_2 8$$
$$\therefore 8 = 2^y$$
$$y = 3$$

x	$y = \log_2 x$
8	$\log_2(8) = \log_2(2^3) = 3$
4	$\log_2(4) = \log_2(2^2) = 2$
2	$\log_2(2) = \log_2(2^1) = 1$
1	$\log_2(1) = \log_2(2^0) = 0$
0	$\log_2(0)$ is undefined
$\dfrac{1}{2}$	$\log_2(\dfrac{1}{2}) = \log_2(2^{-1}) = -1$
$\dfrac{1}{4}$	$\log_2(\dfrac{1}{4}) = \log_2(2^{-2}) = -2$
$\dfrac{1}{8}$	$\log_2(\dfrac{1}{8}) = \log_2(2^{-3}) = -3$

You can use your calculator to work out values for logs only if the base is **10** or **e** (see page 39).

Remember that $\log_{anything} 1 = 0$ because $(anything)^0 = 1$.

Plotting this graph is easiest if you give x the values of powers of 2.

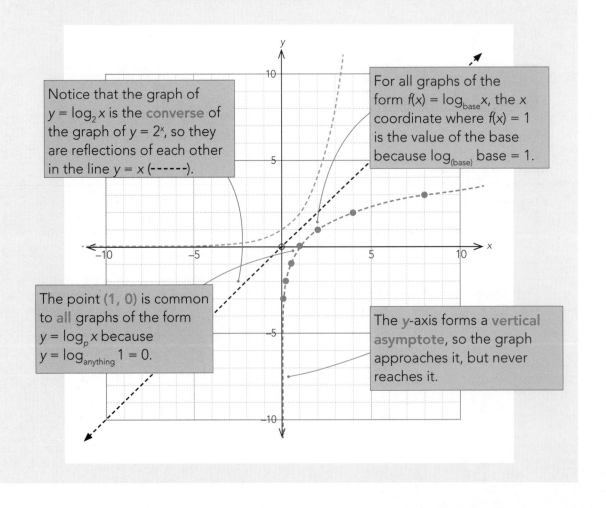

Notice that the graph of $y = \log_2 x$ is the **converse** of the graph of $y = 2^x$, so they are reflections of each other in the line $y = x$ (------).

For all graphs of the form $f(x) = \log_{base}x$, the x coordinate where $f(x) = 1$ is the value of the base because $\log_{(base)} base = 1$.

The point (**1, 0**) is common to **all** graphs of the form $y = \log_p x$ because $y = \log_{anything} 1 = 0$.

The **y-axis** forms a **vertical asymptote**, so the graph approaches it, but never reaches it.

    ISBN: 9780170415989

Note:
- **Natural** logarithms have a base **e** (an irrational number a bit like π, but $e = 2.718\ldots\ldots$), and they are written $\log_e x$ or **ln x**.
- The graph of $f(x) = \log_e x$ is the **converse** of the graph of $f(x) = e^x$.
- On your calculator:

  $\boxed{\log}$ means $\log_{10}$ and $\boxed{\ln}$ means $\log_e$

Complete the following tables and graphs.

5 **a** $y = \log_3 x$

x	$y = \log_3 x$
9	
3	
1	$\log_3(1) = \log_3(3^0) = 0$
0	$\log_3(0)$ is undefined
$\frac{1}{3}$	$\log_3(\frac{1}{3}) = \log_3(3^{-1}) = -1$
$\frac{1}{9}$	

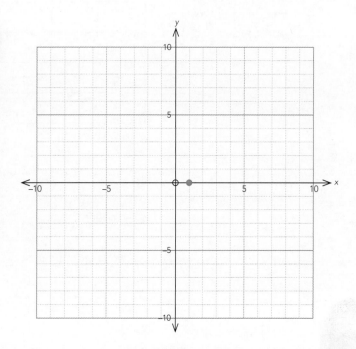

b $y = \log_{10} x$

x	$y = \log_{10} x$
100	
10	
1	
0	
$\frac{1}{10}$	
$\frac{1}{100}$	

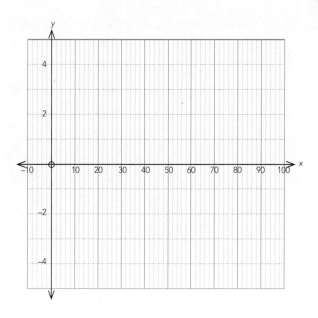

ISBN: 9780170415989

6 Square root graphs

- These are graphs whose equation includes a $\sqrt{x}$.

Example: $f(x) = \sqrt{x}$

x	$f(x) = \sqrt{x}$
16	4
9	3
4	2
1	1
0	0
−1	

Plotting this graph is easiest if you give x the values of square numbers.

Remember that there are no real solutions to $y = \sqrt{\text{negative number}}$

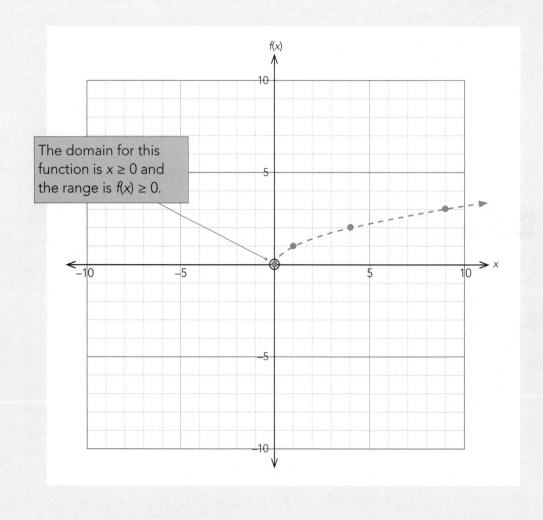

The domain for this function is $x \geq 0$ and the range is $f(x) \geq 0$.

ISBN: 9780170415989

7 Absolute value graphs

- The absolute value of x is written **|x|**.
- The absolute value of a number is its **magnitude**, or how big it is. Example: $|{-2}| = 2$.
- Absolute values are **always positive**.

Example: $f(x) = |x|$

x	f(x) = |x|
3	3
2	2
1	1
0	0
−1	1
−2	2
−3	3

|−2| = 2

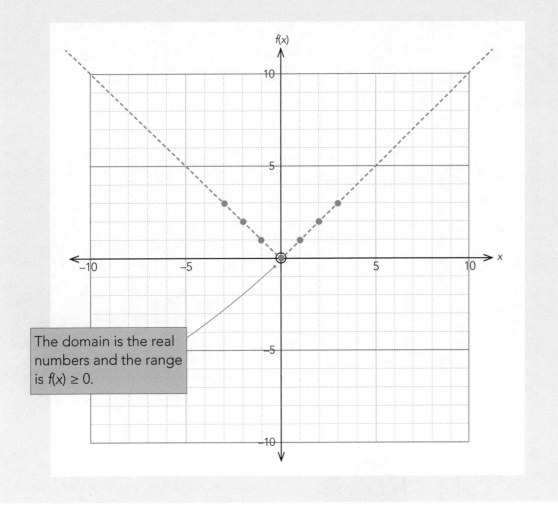

The domain is the real numbers and the range is $f(x) \geq 0$.

ISBN: 9780170415989

8 Trigonometric graphs

- For this standard, these are graphs of equations which include sin(x), cos(x) or tan(x).
- Graphs of trigonometric functions have many practical uses, e.g. in studies of sound and light, tides, etc.
- In order to plot these, make sure your calculator is set on degrees (not radians or gradians).
- Rounding these to 2 dp for plotting is sufficient, but for calculations you should round to 4 dp.

Graph of f(x) = sin(x)

Complete the table below, and use it to complete the graph.

x°	0	20	40	60	80	90	100	120	140	160	180	200	220	240	260	270	280	300	320	340	360	380	400
sin(x)	0	0.34	0.64	0.87	0.98	1	0.98				0					-1					0	0.34	0.64

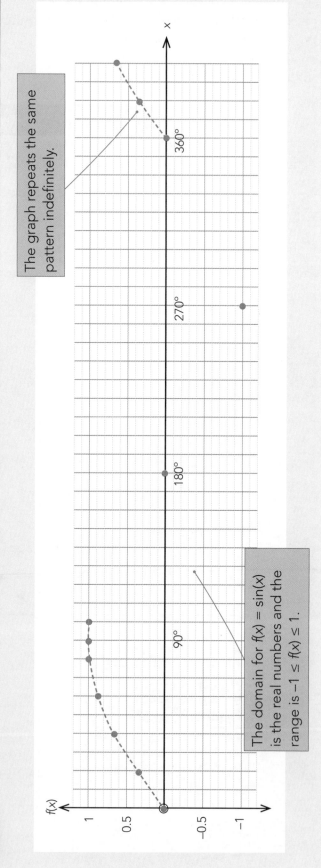

The graph repeats the same pattern indefinitely.

The domain for f(x) = sin(x) is the real numbers and the range is −1 ≤ f(x) ≤ 1.

ISBN: 9780170415989

Graph of $f(x) = \cos(x)$

Complete the table below, and use it to complete the graph.

x°	0	20	40	60	80	90	100	120	140	160	180	200	220	240	260	270	280	300	320	340	360	380	400
cos(x)	1	0.94	0.77	0.5	0.17	0	-0.17	-0.5			-1					0					1	0.94	0.77

This graph also repeats the same pattern indefinitely.

Graph of $y = \tan(x)$

Complete the table below, and use it to complete the graph.

x°	0	20	40	60	80	90	100	120	140	160	180	200	220	240	260	270	280	300	320	340	360	380	400
tan(x)	0	0.36	0.84	1.73	5.67	U	−5.67	−1.73			0					U					0	0.36	0.84

Undefined

There are vertical asymptotes at x = 90°, 270°, 450°, 630°, etc., which means that tan(x) is undefined at these points.

This graph also repeats the same pattern indefinitely.

Properties of functions

Properties of functions apply to **all** functions within a group.

Example 1: All cubics have **rotational symmetry**.

$f(x) = (x + 1)(x - 2)^2$ has rotational symmetry of order 2, with a centre of rotation at (1, 2).

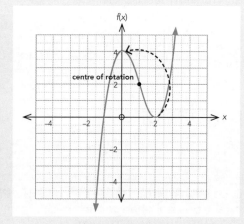

Example 2: All absolute value graphs have **reflective symmetry**.

$f(x) = |x - 2| + 3$ has reflective symmetry in the line $x = 2$.

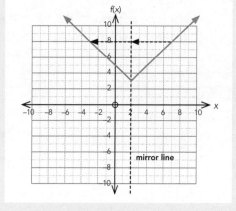

Example 3: All trigonometric functions are **periodic**. This means that the graph repeats the same pattern indefinitely.

$f(x) = \sin(x)$ is periodic, with a period of 360° and an amplitude of 1.

Note: It also has
- **rotational symmetry** around the points (−360°, 0), (0°, 0), (360°, 0), etc.
- **reflective symmetry** with mirror lines at $x = -270°$, −90°, 90°, 270°, etc.

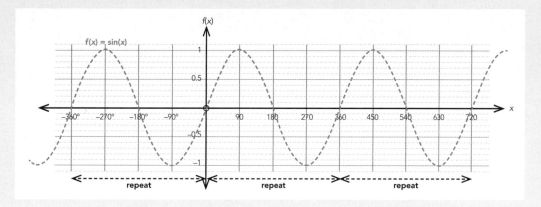

ISBN: 9780170415989

Transformations of graphs

- You need to be able to transform any of the graphs we have considered so far.
- Transformations include:
  - translations: vertical, horizontal and combinations of these
  - reflections: in the x and y axes
  - enlargements: vertical and horizontal.

Translations

1 Vertical translations

- A vertical translation shifts the graph **up** or **down**.
- It does not change the shape.
- You can shift a graph
  - **up** by **adding** a number to the function
  - or **down** by **subtracting** a number from a function.

Examples:

1 $f(x) = |x| - 4$

$-4 \Rightarrow$ **down 4**

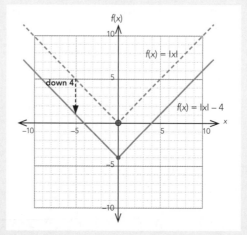

2 $f(x) = \sin(x) + 0.5$

$+0.5 \Rightarrow$ **up 0.5**

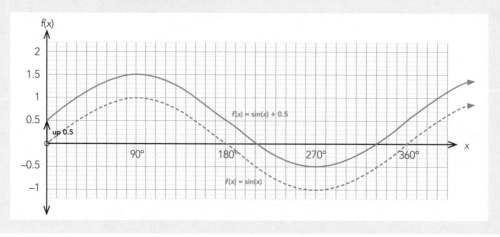

Draw the following functions, and write down the domain and range for each. Draw and label asymptotes on those that have them. Describe each transformation.

1 $f(x) = x^3 + 3$

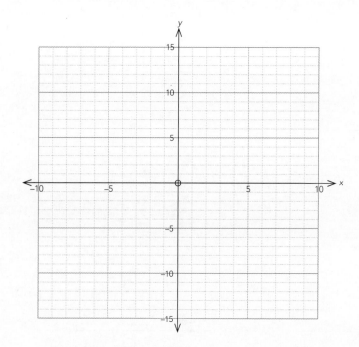

Domain: _____

Range: _____

Transformation from $f(x) = x^3$: _____

2 $f(x) = \cos(x) - 0.5$

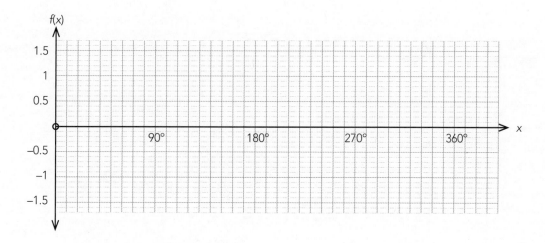

Domain: _____

Range: _____

Transformation from $f(x) = \cos(x)$: _____

3 $f(x) = \left(\frac{1}{2}\right)^x - 3$

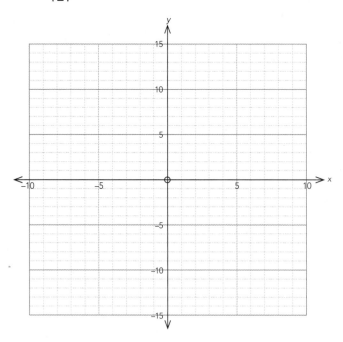

Transformation from $f(x) = \left(\frac{1}{2}\right)^x$: _____

4 $y = \log_2 x + 1$

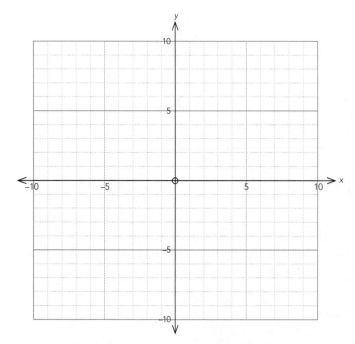

Domain: _____

Range: _____

Transformation from $y = \log_2 x$: _____

2 Horizontal translations

- A horizontal translation shifts the graph **left** or **right**.
- It does not change the shape.
- You can shift a graph
    - **left** by **adding** a number to the x term
    - or   **right** by **subtracting** a number from the x term.

> Notice that the graphs move in the **opposite** direction to what you might expect.

Examples:

1 $f(x) = \sqrt{x + 3}$

> Be careful — this is **not** the same as $f(x) = \sqrt{x} + 3$.

$+ 3 \Rightarrow$ **left 3**

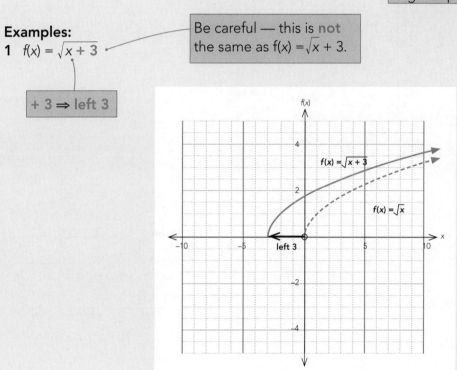

2 $f(x) = \cos(x - 90°)$

> Be careful — this is **not** the same as $\cos(x) - 90°$.

$- 90° \Rightarrow$ **right 90°**

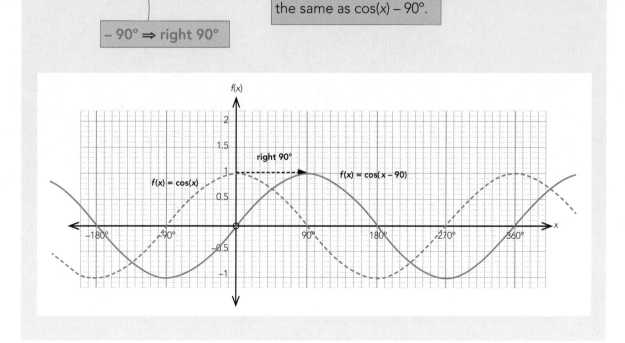

Draw the following functions, and write down the domain and range for each. Draw and label asymptotes on those that have them. Describe each transformation.

1 $y = 3^{(x + 2)}$

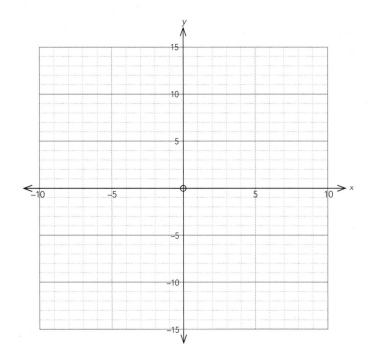

Domain: _____

Range: _____

Transformation from $y = 3^x$: _____

2 $(x - 5)y = 12$

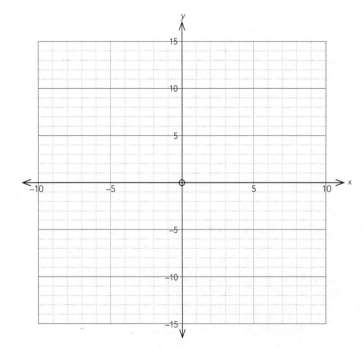

Domain: _____

Range: _____

Transformation from $xy = 12$: _____

3 $f(x) = \sqrt{x + 2}$

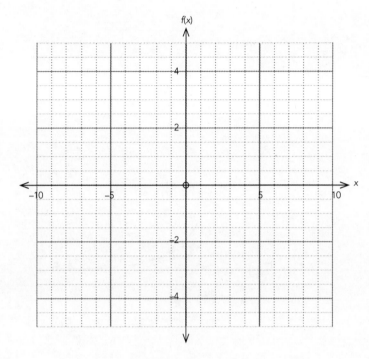

Domain: _____

Range: _____

Transformation from $f(x) = \sqrt{x}$: _____

4 $f(x) = \sin(x + 45°)$

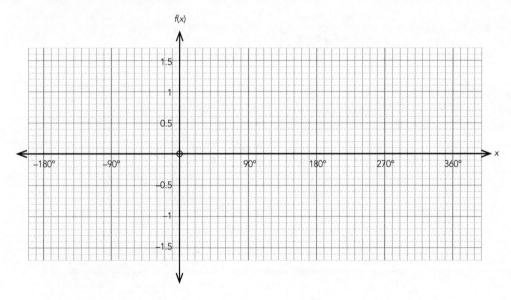

Domain: _____

Range: _____

Transformation from $f(x) = \sin(x)$: _____

3 Combinations of vertical and horizontal translations

Examples:

1 $f(x) = \left(\frac{1}{2}\right)^{x-4} + 3$

$+3 \Rightarrow$ up 3 $\Rightarrow$
asymptote is at $f(x) = 3$.

$-4 \Rightarrow$ right 4

Remember, (0, 1) is a critical point. **All exponential graphs in their basic form pass through this. It is always one unit above or below the asymptote.**

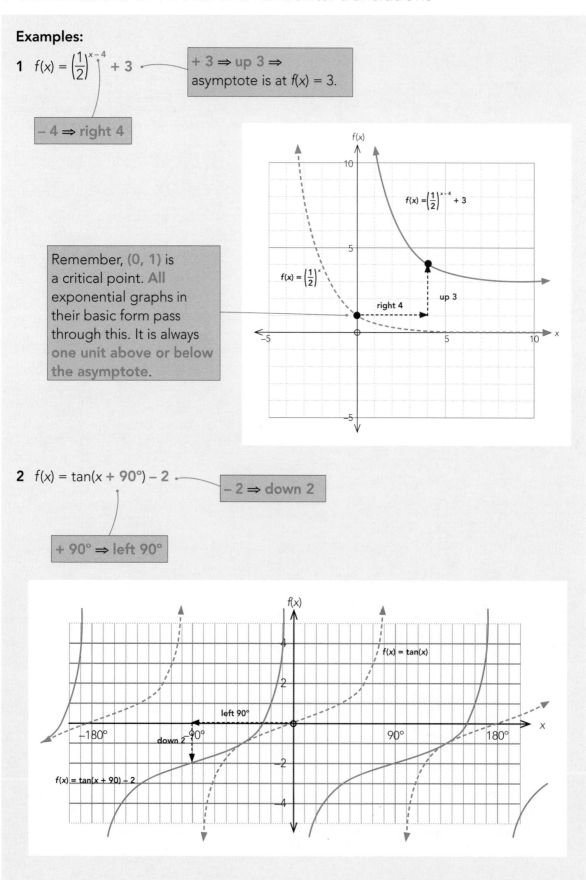

2 $f(x) = \tan(x + 90°) - 2$

$-2 \Rightarrow$ down 2

$+90° \Rightarrow$ left 90°

Draw the following functions, and write down the domain and range for each. Draw and label asymptotes on those that have them. Describe each transformation.

1 $f(x) = (x - 2)^4 + 3$

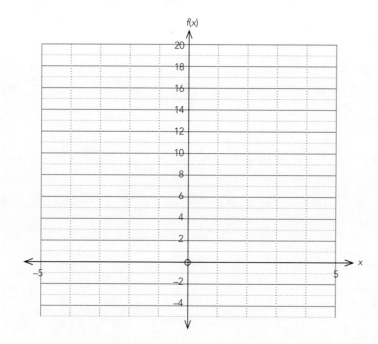

Domain: _____

Range: _____

Transformation from $f(x) = x^4$: _____

2 $y = 2^{(x-3)} + 1$

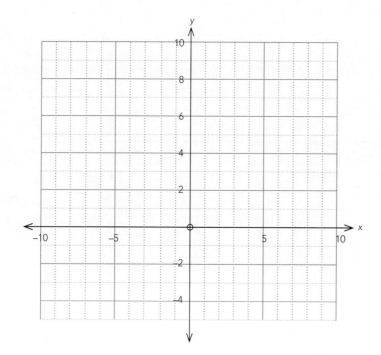

Domain: _____

Range: _____

Transformation from $y = 2^x$: _____

3 $f(x) = |(x - 1)| + 4$

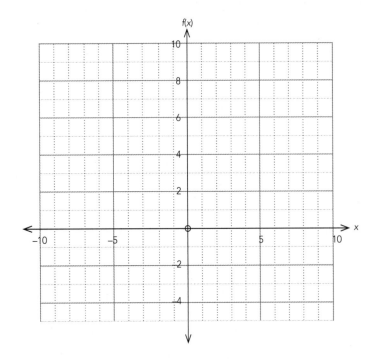

Domain: _____

Range: _____

Transformation from $f(x) = |x|$: _____

4 $f(x) = \cos(x + 45°) + 3$

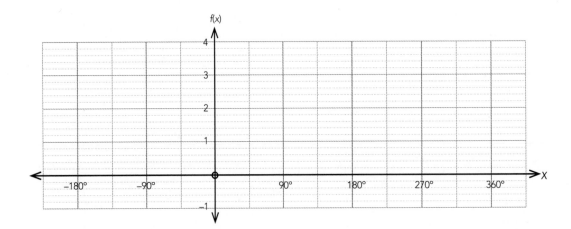

Domain: _____

Range: _____

Transformation from $f(x) = \cos(x)$: _____

Mix and match

Match each graph with its equation.

1

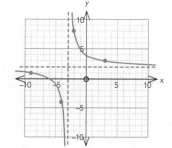

_____

2

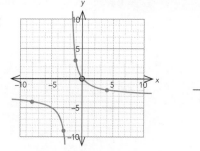

_____

3

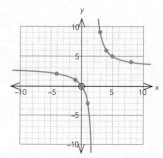

_____

4

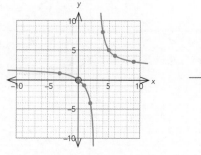

_____

A $y = \dfrac{6}{x+3} + 2$ **B** $y = \dfrac{6}{x-2} + 3$ **C** $y = \dfrac{6}{x-3} + 2$ **D** $y = \dfrac{6}{x+2} - 3$

5

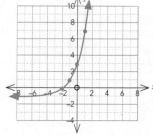

_____

6

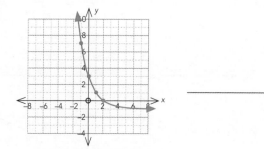

_____

7

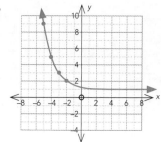

_____

8

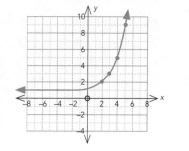

_____

A $y = \left(\dfrac{1}{2}\right)^{x-2} - 1$ **B** $y = \left(\dfrac{1}{2}\right)^{x+2} + 1$ **C** $y = 2^{x-2} + 1$ **D** $y = 2^{x+2} - 1$

Hint for 9–13: Consider where the point (0, 0) on $f(x) = \sin(x)$ has moved to.

9

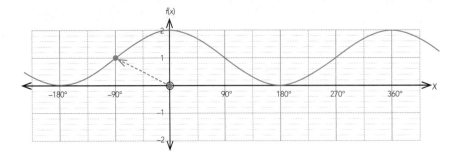

_____

10

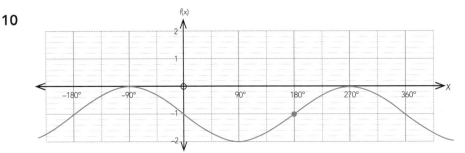

_____

11

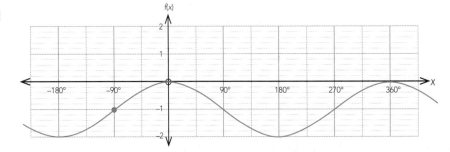

_____

12

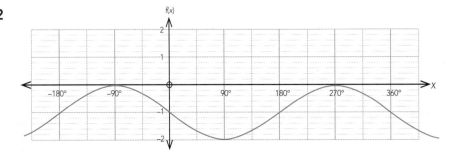

_____

13

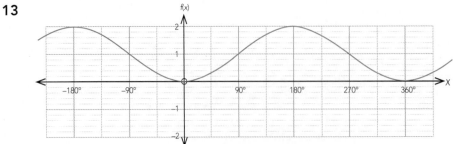

_____

A $f(x) = \sin(x + 180°) - 1$ **B** $f(x) = \sin(x - 180°) - 1$

C $f(x) = \sin(x - 90°) + 1$ **D** $f(x) = \sin(x + 90°) + 1$

E $f(x) = \sin(x + 90°) - 1$

 ISBN: 9780170415989

Write equations for the following graphs.

14

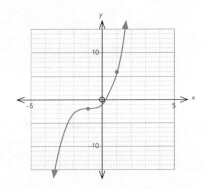

15

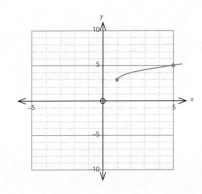

16

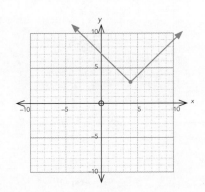

17

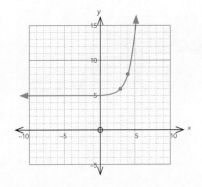

18 This graph is a function of $f(x) = \cos(x)$.

Hint: Consider where the point $(0, 1)$ on $f(x) = \cos(x)$ has moved to.

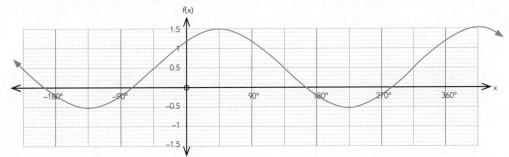

19 This graph is a function of $f(x) = \tan(x)$.

Hint: Consider where the point $(0, 0)$ on $f(x) = \tan(x)$ has moved to.

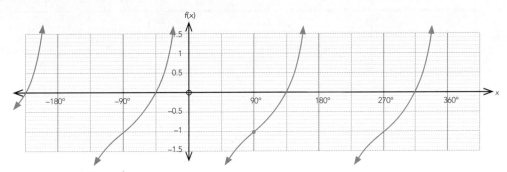

Reflections

1 Reflections in the x-axis

- The **x-axis** becomes the **mirror line**.
- This does not change the shape.
- You can reflect a function in the x-axis by **changing the sign of the entire function**.

Examples:

1 $f(x) = -(|x| + 1)$

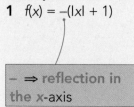
$-\ \Rightarrow$ **reflection in** the x-axis

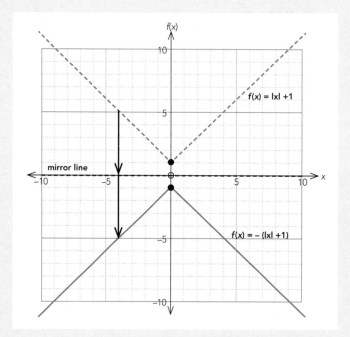

2 $f(x) = -\sin(x)$

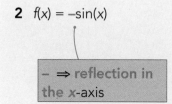
$-\ \Rightarrow$ **reflection in** the x-axis

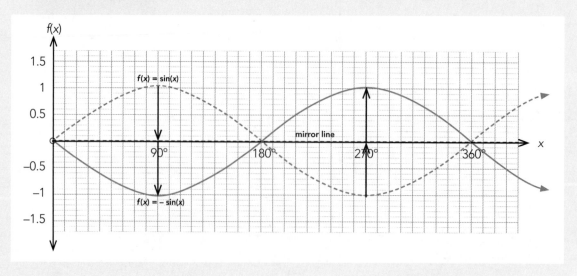

Draw the following functions, and write down the domain and range for each. Draw and label asymptotes on those that have them. Describe each transformation.

1 $f(x) = -x^3$

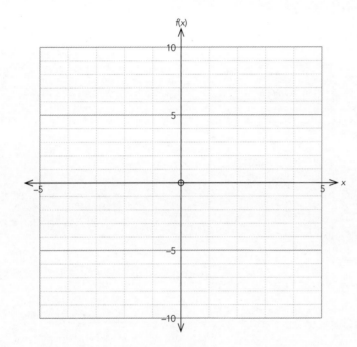

Domain: _____

Range: _____

Transformation from $f(x) = x^3$: _____

2 $y = -\dfrac{8}{x}$

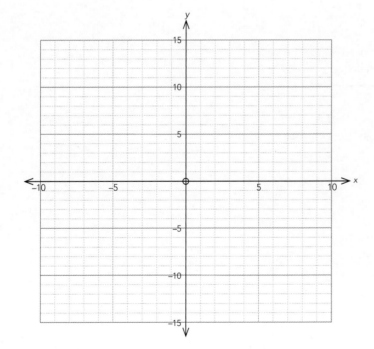

Domain: _____

Range: _____

Transformation from $y = \dfrac{8}{x}$: _____

3 $f(x) = -\log_2 x$

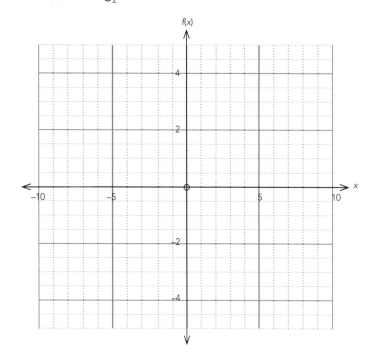

Domain: _____

Range: _____

Transformation from $f(x) = \log_2 x$: _____

4 $f(x) = -\tan(x)$

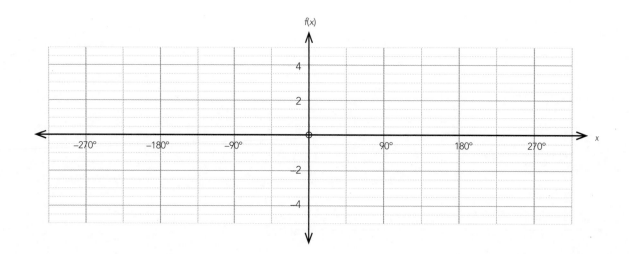

Domain: _____

Range: _____

Transformation from $f(x) = \tan(x)$: _____

2 Reflections in the *y*-axis (or *f*(*x*) axis)

- The **y-axis** becomes the **mirror line**.
- This does not change the shape.
- You can reflect a function in the x-axis by **changing the sign of the x term**.
- This **cannot** be done for $y = \log(x)$ or $y = \sqrt{x}$ because log(negative number) and $\sqrt{\text{negative number}}$ do not exist.
- For some other functions, $f(-x)$ is the same as $-f(x)$.
  For example: $f(x) = (-x)^3$ and $f(x) = -x^3$

$$y = \frac{6}{(-x)} \text{ and } y = -\frac{6}{x}$$

Examples:

1 $f(x) = 2^{(-x)}$

$- \Rightarrow$ reflection in the *y*-axis

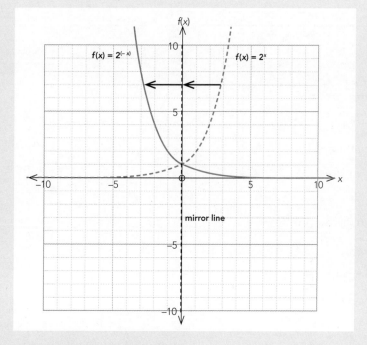

2 $f(x) = \tan(-x)$

$- \Rightarrow$ reflection in the *y*-axis

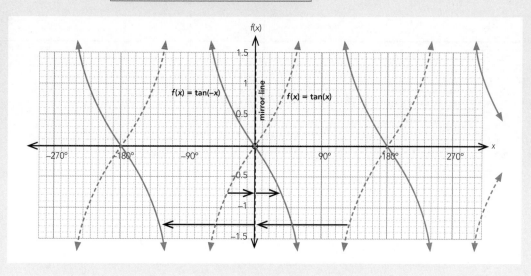

Draw the following functions, and write down the domain and range for each. Draw and label asymptotes on those that have them. Describe each transformation.

1 $y = \left(\frac{1}{2}\right)^{-x}$

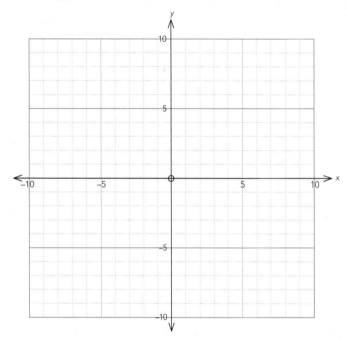

Domain: _____

Range: _____

Transformation from $y = \left(\frac{1}{2}\right)^{x}$: _____

2 $f(x) = \sin(-x)$

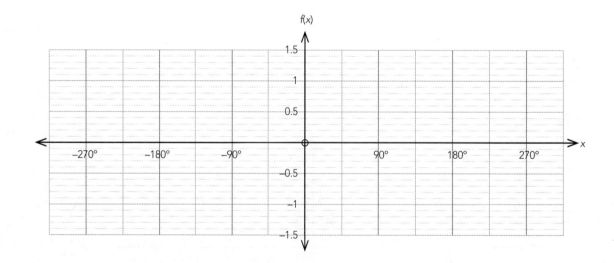

Domain: _____

Range: _____

Transformation from $f(x) = \sin(x)$: _____

Combinations of translations and reflections

Match each graph with its equation.

1

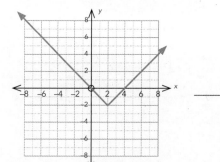 _____

2

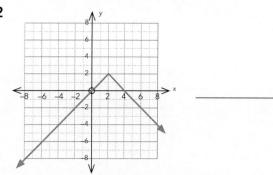 _____

3

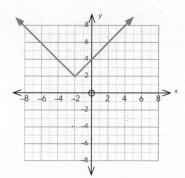 _____

4

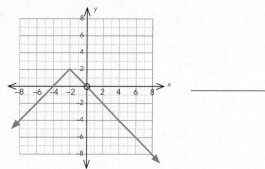 _____

A $y = |x + 2| + 2$ **B** $y = |x - 2| - 2$ **C** $y = 2 - |x - 2|$ **D** $y = 2 - |x + 2|$

5

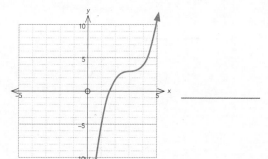 _____

6

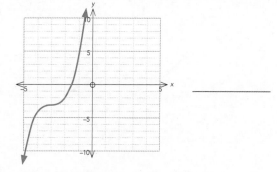 _____

7

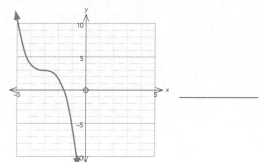 _____

8

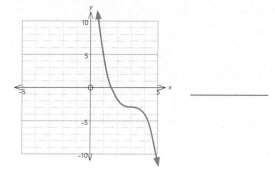 _____

A $y = (3 - x)^3 - 3$ **B** $y = 3 - (x + 3)^3$ **C** $y = (x - 3)^3 + 3$ **D** $y = (x + 3)^3 - 3$

9

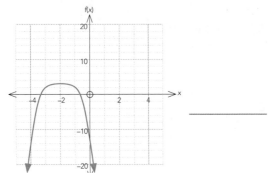

_____

10

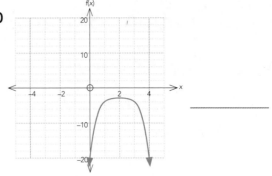

_____

11

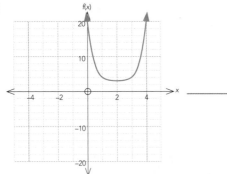

_____

12

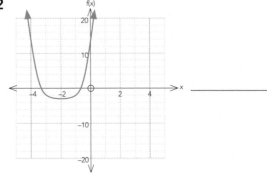

_____

A $f(x) = -(x + 2)^4 + 3$ **B** $f(x) = (x + 2)^4 - 3$ **C** $f(x) = -(x - 2)^4 - 3$ **D** $f(x) = (x - 2)^4 + 3$

13

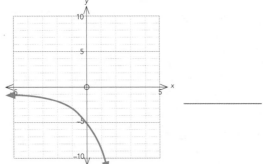

_____

14

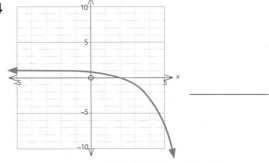

_____

15

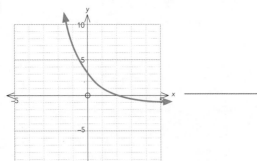

_____

16

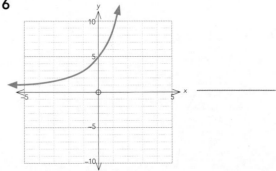

_____

A $y = 2^{(2-x)} - 1$ **B** $y = -2^{(x+2)} - 1$ **C** $y = 2^{(x+2)} + 1$ **D** $y = -2^{(x-2)} + 1$

Write equations for the following graphs.

17

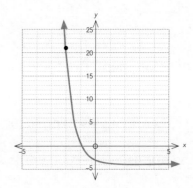

18

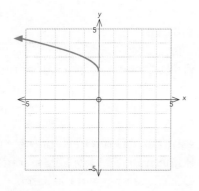

---

19 This graph is a function of $f(x) = \tan(x)$.
Hint: Consider where the point $(0, 0)$ on $f(x) = \tan(x)$ has moved to.

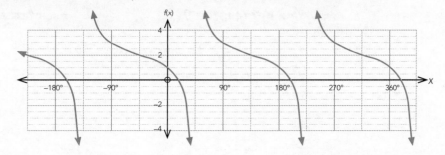

---

20 This graph is a function of $f(x) = \cos(x)$.
Hint: Consider where the point $(0, 1)$ on $f(x) = \cos(x)$ has moved to.

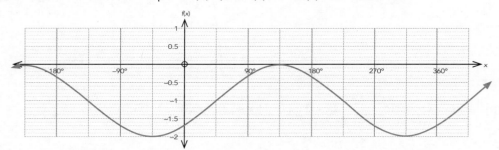

---

21 This graph is a function of $f(x) = \sin(x)$.
Hint: Consider where the point $(0, 0)$ on $f(x) = \sin(x)$ has moved to.

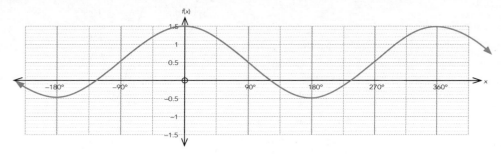

More on cubic and quartic graphs

Translations

- While the equation $y = x^3$ appears to have just **one x intercept**, this could be considered to be **three identical intercepts**.
- $y = x^3$ could also be written as $y = (x + 0)(x + 0)(x + 0)$.
- Each of these intercepts can be translated to the left or right by adding to or subtracting from each of the x factors.
- Remember that horizontal translations move in the **opposite** direction to the sign.
- Quartics behave in the same way.
- The y or f(x) intercept can be calculated by multiplying out the **constants**.

Example 1: Cubic: $f(x) = x(x - 2)(x + 3)$ or $f(x) = (x + 0)(x - 2)(x + 3)$

+ 0 ⇒ x intercept at 0

+ 3 ⇒ x intercept at –3

– 2 ⇒ x intercept at +2

$f(x)$ intercept = **+0** x **–2** x **+3** = 0

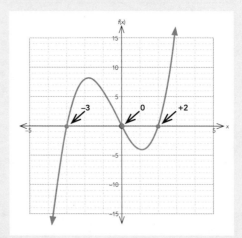

Example 2: Quartic: $f(x) = (x - 1)(x + 1)(x + 2)(x - 3)$

$f(x)$ intercept = **–1** x **+1** x **+2** x **–3** = +6

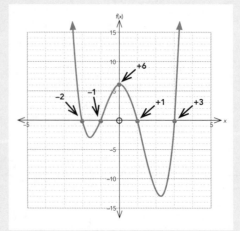

  ISBN: 9780170415989

Reflections in the *x*-axis

These can be achieved in two ways:

1 A negative sign in front of the **entire** function, e.g. $y = -x(x - 2)(x + 3)$.

2 Reversing the order within a bracket containing a negative sign,
e.g. $y = x(2 - x)(x + 3)$.

Remember:

$y = x^3$ $y = -x^3$ $y = x^4$ $y = -x^4$

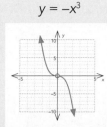 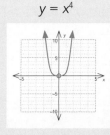 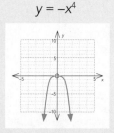

You should check the sign of the x^3 or x^4 term by multiplying the x terms only. The 'arms' of the curve will go in the same directions as in these basic graphs.

1 $y = -x(x - 2)(x + 3)$ Sign of $x^3 = -x$ times x times $x = -x^3 \Rightarrow$

     y intercept $= 0 \times -2 \times +3 = 0$

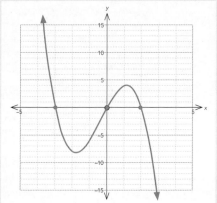

2 $y = (x - 1)(x + 1)(x + 2)(3 - x)$

     Sign of $x^4 = x$ times x times x times $-x = -x^4 \Rightarrow$

     y intercept $= -1 \times +1 \times +2 \times +3 = -6$

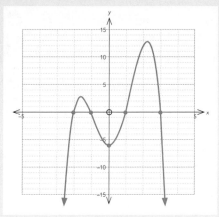

Match each graph with its equation.

1

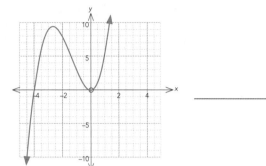

_____

2

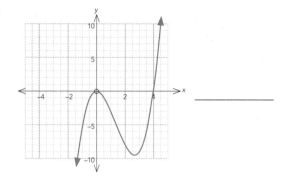

_____

3

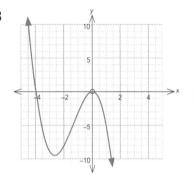

_____

4

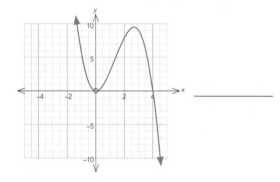

_____

A $y = -x^2(x - 4)$ **B** $y = x^2(x + 4)$ **C** $y = -x^2(4 - x)$ **D** $y = -x^2(4 + x)$

5

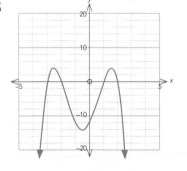

_____

6

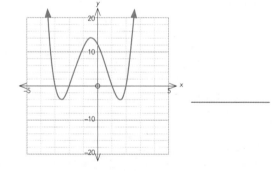

_____

7

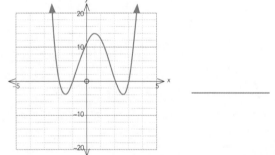

_____

8

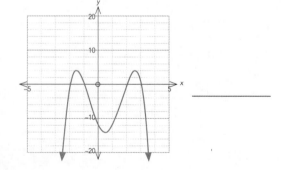

_____

A $y = -(x + 2)(x - 1)(x - 2)(x + 3)$ **B** $y = -(x + 2)(x + 1)(2 - x)(x - 3)$

C $y = -(x + 2)(x + 1)(x - 2)(x - 3)$ **D** $y = (x + 2)(x - 1)(x - 2)(x + 3)$

Write equations for the following graphs.

9

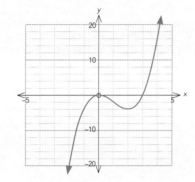

_____

10

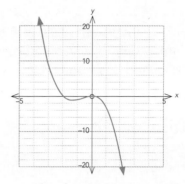

_____

11

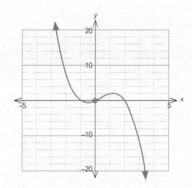

_____

12

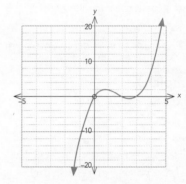

_____

13

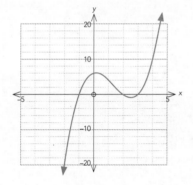

_____

14

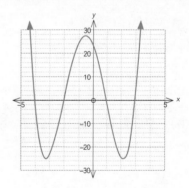

_____

15

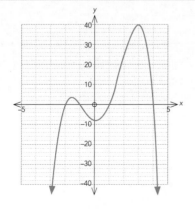

_____

16

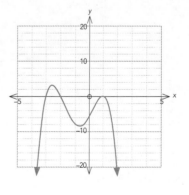

_____

Enlargements

- Enlargements can occur in a **vertical** or **horizontal** direction.
- Remember, 'enlargement' is a general term, which may result in the curve being **stretched** or **shrunk** in one direction.

1 Vertical enlargements

- These cause stretching or shortening along the y or $f(x)$ axis.
- They **do** change the shape.
- You can enlarge a function along the y or $f(x)$ axis by **multiplying the entire function by a number (a)**.

$a > 1 \rightarrow$ function is **stretched** vertically ↕ $a < 1 \rightarrow$ function is **shortened** vertically ↕

Examples:

1 $f(x) = 2|x|$

> $2 \Rightarrow$ **stretch** along the $f(x)$ **axis**. The $f(x)$ values are **doubled**.

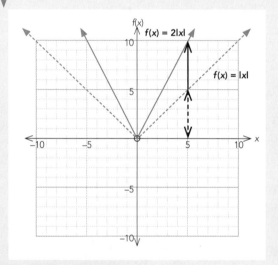

2 $f(x) = 0.5\sin(x)$

> $0.5 \Rightarrow$ **shortening** along the $f(x)$ axis. The $f(x)$ values are **halved**.

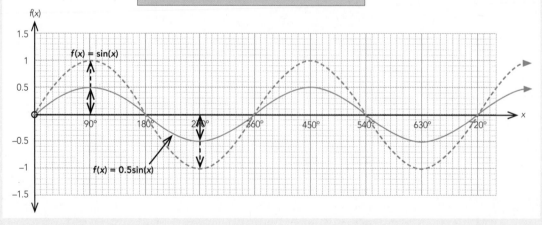

Period = 360°
Amplitude = 0.5

Draw the following functions, and write down the domain and range for each. Draw and label asymptotes on those that have them. Describe each transformation.

1 $f(x) = 2x^3$

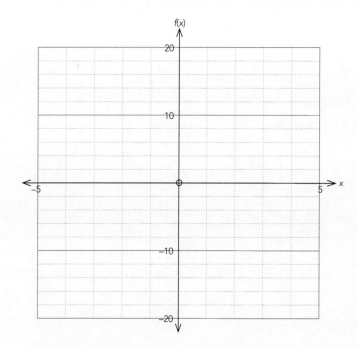

Domain: _____

Range: _____

Transformation from $f(x) = x^3$: _____

2 $y = \dfrac{1}{2}(4^x)$

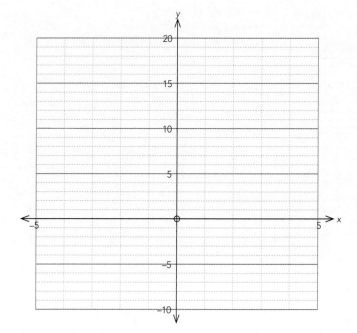

Domain: _____

Range: _____

Transformation from $y = 4^x$: _____

3 $f(x) = 3 \log_2 x$

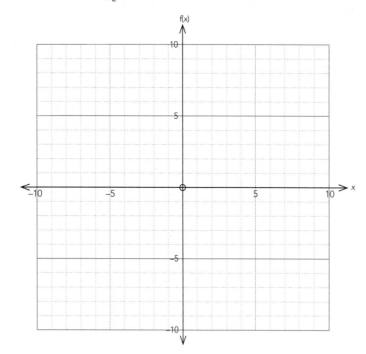

Domain: _____

Range: _____

Transformation from $f(x) = \log_2 x$: _____

4 $f(x) = 5\cos(x)$

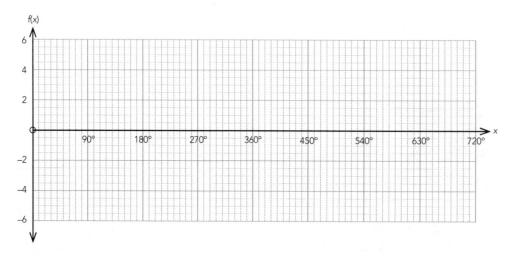

Domain: _____

Range: _____

Period: _____

Amplitude: _____

Transformation from $f(x) = \cos(x)$: _____

2 Horizontal enlargements

- These cause **stretching** or **shortening** along the **x**-axis.
- They **do** change the shape.
- They are particularly important for **trigonometric functions**, because they change the period.
- You can enlarge a function along the x-axis by **multiplying x by a number (a)**.

a > 1 → function is **shortened** horizontally →◄

a < 1 → function is **stretched** horizontally ◄►

Notice that this is the opposite of the changes for vertical enlargements.

1 $f(x) = |2x|$

2 ⇒ **shortening** along the **x-axis**. The x values are **doubled**.

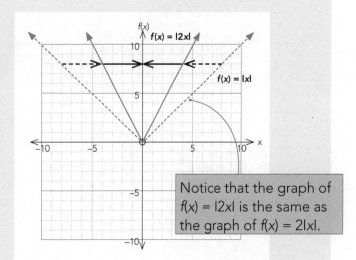

Notice that the graph of $f(x) = |2x|$ is the same as the graph of $f(x) = 2|x|$.

2 $f(x) = \sin(0.5x)$

0.5 ⇒ **stretching** along the x-axis. The x values are **halved**.

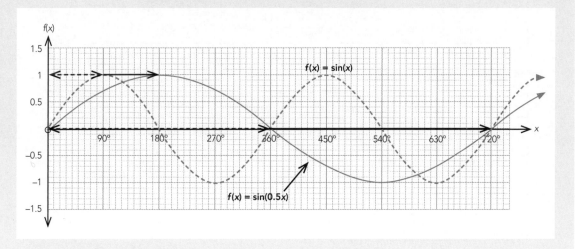

Period = 720°
Amplitude = 1

Draw the following functions, and write down the domain and range for each. Draw and label asymptotes on those that have them. Describe each transformation.

1 $y = 3^{2x}$

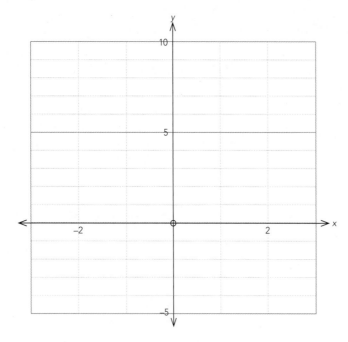

Domain: _____

Range: _____

Transformation from $y = 3^x$: _____

2 $f(x) = \cos(3x)$

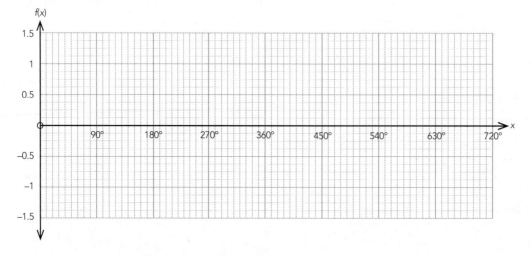

Domain: _____

Range: _____

Period: _____

Amplitude: _____

Transformation from $f(x) = \cos(x)$: _____

3 $y = (2x)^2$

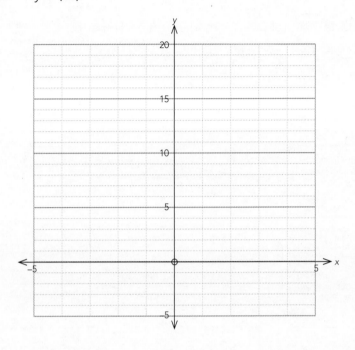

Domain: _____

Range: _____

Transformation from $y = x^2$: _____

4 $f(x) = \tan(0.5x)$

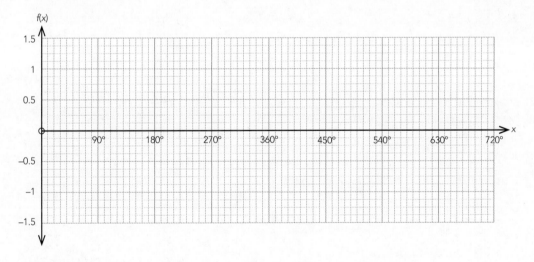

Domain: _____

Range: _____

Period: _____

Amplitude: _____

Transformation from $f(x) = \tan(x)$: _____

3 Combinations of vertical and horizontal enlargements

Match each graph with its equation. One integral point is marked on each graph.

1

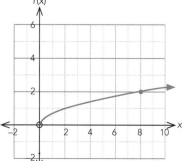

_____

2

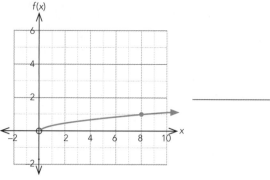

_____

3

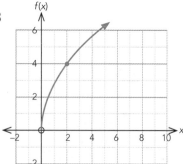

_____

4

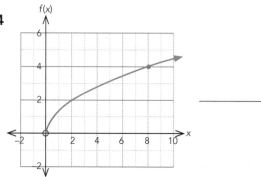

_____

A $f(x) = 2\sqrt{0.5x}$ **B** $f(x) = 0.5\sqrt{0.5x}$ **C** $f(x) = 0.5\sqrt{2x}$ **D** $f(x) = 2\sqrt{2x}$

5

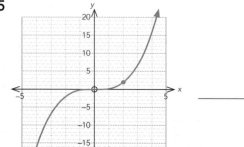

_____

6

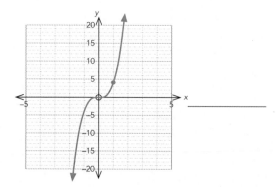

_____

7

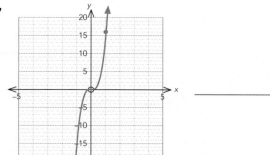

_____

8

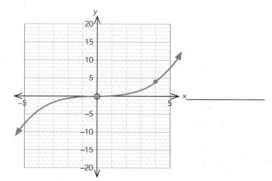

_____

A $y = 2(0.5x)^3$ **B** $y = 0.5(0.5x)^3$ **C** $y = 2(2x)^3$ **D** $y = 0.5(2x)^3$

Hint for 9–13: Using $f(x) = \sin(x)$ as a basis, compare the period and the amplitude.

9

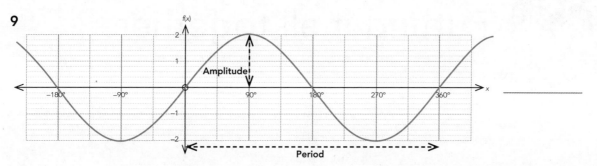

10

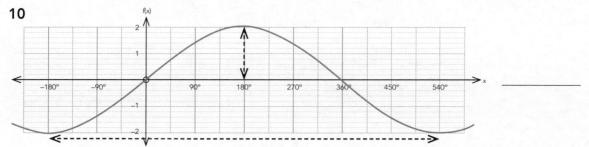

11

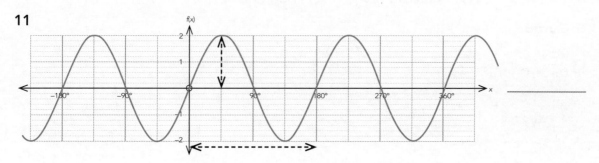

12

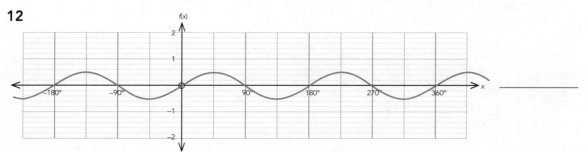

13

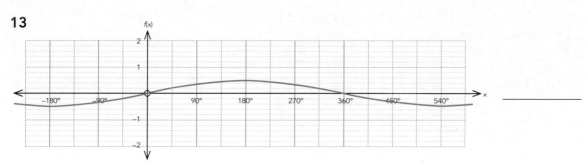

A $f(x) = 2\sin(0.5x)$ **B** $f(x) = 0.5\sin(2x)$

C $f(x) = 0.5\sin(0.5x)$ **D** $f(x) = 2\sin(x)$

E $f(x) = 2\sin(2x)$

Putting it all together

In general:

$a > 1 \Rightarrow$ stretched

$a < 1 \Rightarrow$ shortened

$+ c \Rightarrow$ shift up

$- c \Rightarrow$ shift down

$$f(x) = a(b \times \text{function} \pm d) \pm c$$

$b > 1 \Rightarrow$ shortened

$b < 1 \Rightarrow$ stretched

$+ d \Rightarrow$ shift left

$- d \Rightarrow$ shift right

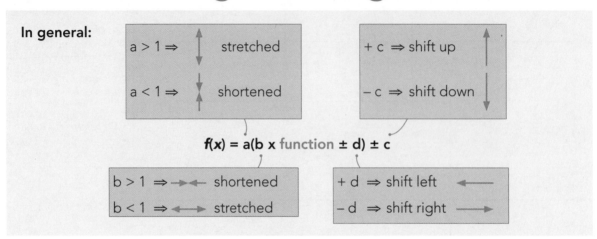

Summary for each type of graph

Parabolas

Intercept form

$+ a \Rightarrow$ $- a \Rightarrow$

$$f(x) = a(x - p)(x + q)$$

a stretches or shortens the graph vertically ∴ alters the $f(x)$ intercept.

x intercepts at (p, 0) and (−q, 0).

Axis of symmetry halfway between intercepts

(−q, 0)   (p, 0)

Minimum or maximum point on the axis of symmetry

$f(x)$ intercept = apq

Turning point form

$+ a \Rightarrow$ $- a \Rightarrow$

$$f(x) = a(x - d)^2 + c$$

a stretches or shortens the graph vertically ∴ alters the $f(x)$ intercept.

$+ d \Rightarrow$ shift left
$- d \Rightarrow$ shift right
$+ c \Rightarrow$ shift up
$- c \Rightarrow$ shift down

Turning point: (+d, +c)

x intercepts where $f(x) = 0$

Axis of symmetry through turning point

$f(x)$ intercept = $ab^2 + c$

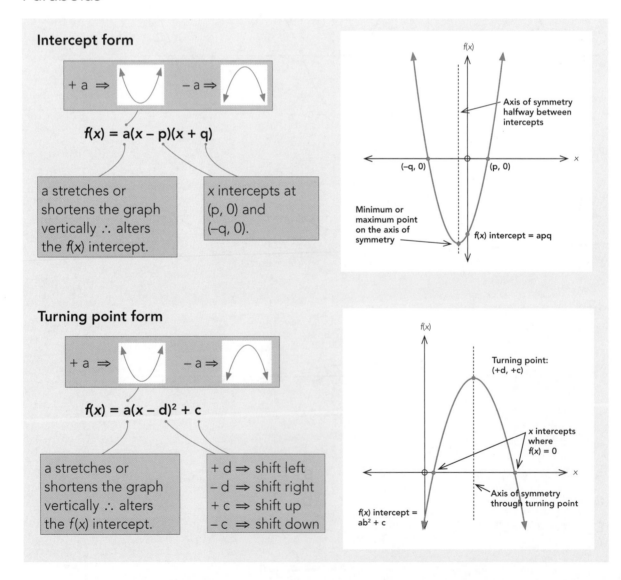

 ISBN: 9780170415989

Cubics

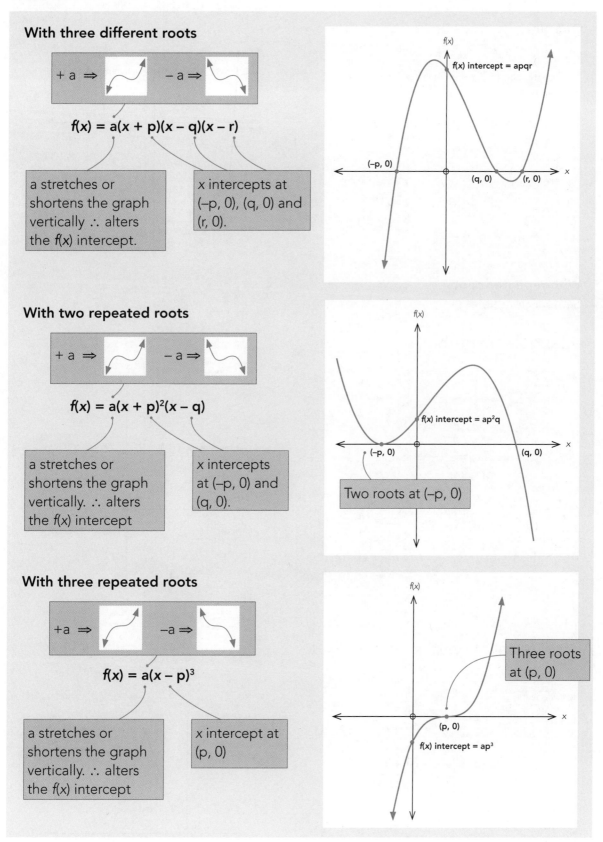

With three different roots

$+a \Rightarrow$ $-a \Rightarrow$

$$f(x) = a(x + p)(x - q)(x - r)$$

a stretches or shortens the graph vertically ∴ alters the $f(x)$ intercept.

x intercepts at $(-p, 0)$, $(q, 0)$ and $(r, 0)$.

$f(x)$ intercept = apqr

$(-p, 0)$ $(q, 0)$ $(r, 0)$

With two repeated roots

$+a \Rightarrow$ $-a \Rightarrow$

$$f(x) = a(x + p)^2(x - q)$$

a stretches or shortens the graph vertically. ∴ alters the $f(x)$ intercept

x intercepts at $(-p, 0)$ and $(q, 0)$.

$f(x)$ intercept = ap^2q

$(-p, 0)$ $(q, 0)$

Two roots at $(-p, 0)$

With three repeated roots

$+a \Rightarrow$ $-a \Rightarrow$

$$f(x) = a(x - p)^3$$

a stretches or shortens the graph vertically. ∴ alters the $f(x)$ intercept

x intercept at $(p, 0)$

Three roots at $(p, 0)$

$(p, 0)$

$f(x)$ intercept = ap^3

Quartics

Quartics

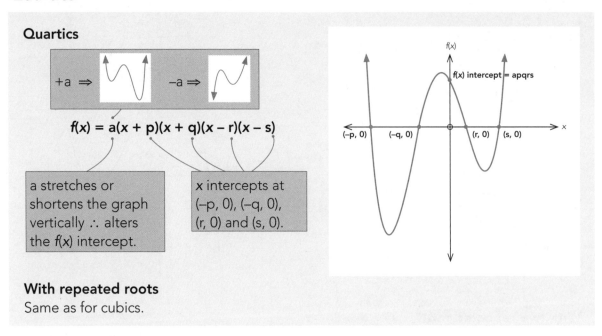

$f(x) = a(x + p)(x + q)(x − r)(x − s)$

a stretches or shortens the graph vertically ∴ alters the $f(x)$ intercept.

x intercepts at (–p, 0), (–q, 0), (r, 0) and (s, 0).

With repeated roots
Same as for cubics.

Rectangular hyperbolae

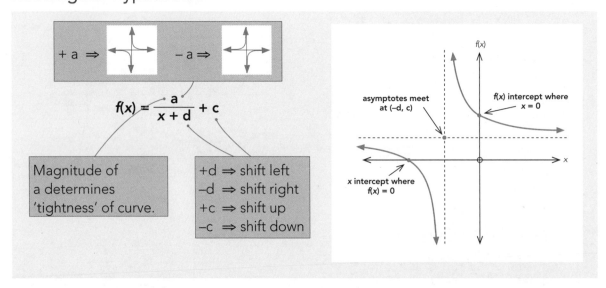

$f(x) = \dfrac{a}{x + d} + c$

Magnitude of a determines 'tightness' of curve.

+d ⇒ shift left
–d ⇒ shift right
+c ⇒ shift up
–c ⇒ shift down

Absolute value graphs

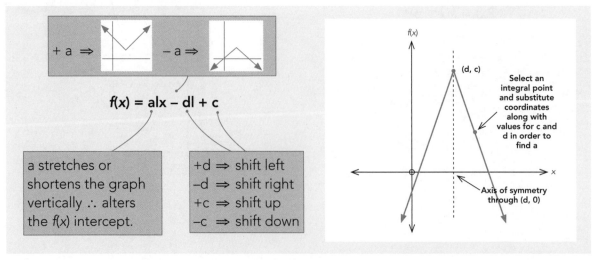

$f(x) = a|x − d| + c$

a stretches or shortens the graph vertically ∴ alters the $f(x)$ intercept.

+d ⇒ shift left
–d ⇒ shift right
+c ⇒ shift up
–c ⇒ shift down

 ISBN: 9780170415989

Exponential graphs

Compare with $f(x) = (\text{base})^x$

With no enlargement

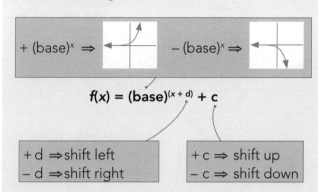

$+ (\text{base})^x \Rightarrow$ $- (\text{base})^x \Rightarrow$

$$f(x) = (\text{base})^{(x + d)} + c$$

| $+d \Rightarrow$ shift left |
| $-d \Rightarrow$ shift right |

| $+c \Rightarrow$ shift up |
| $-c \Rightarrow$ shift down |

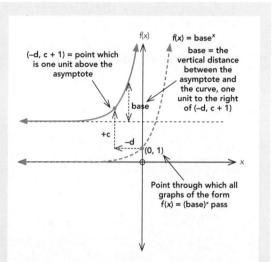

$(-d, c + 1) =$ point which is one unit above the asymptote

$f(x) = \text{base}^x$

base = the vertical distance between the asymptote and the curve, one unit to the right of $(-d, c + 1)$

base

$+c$ $-d$

$(0, 1)$

Point through which all graphs of the form $f(x) = (\text{base})^x$ pass

With enlargement horizontally

$$f(x) = (\text{base})^{bx}$$

b stretches or shortens the graph horizontally.

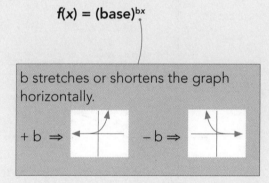

$+b \Rightarrow$ $-b \Rightarrow$

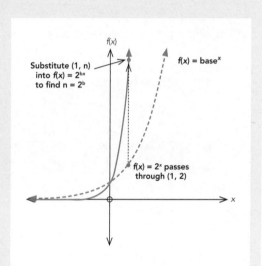

Substitute $(1, n)$ into $f(x) = 2^{bx}$ to find $n = 2^b$

$f(x) = \text{base}^x$

$f(x) = 2^x$ passes through $(1, 2)$

With enlargement vertically

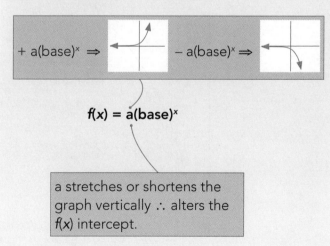

$+ a(\text{base})^x \Rightarrow$ $- a(\text{base})^x \Rightarrow$

$$f(x) = a(\text{base})^x$$

a stretches or shortens the graph vertically $\therefore$ alters the $f(x)$ intercept.

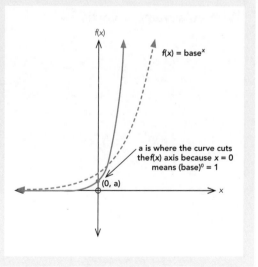

$f(x) = \text{base}^x$

a is where the curve cuts the $f(x)$ axis because $x = 0$ means $(\text{base})^0 = 1$

$(0, a)$

ISBN: 9780170415989

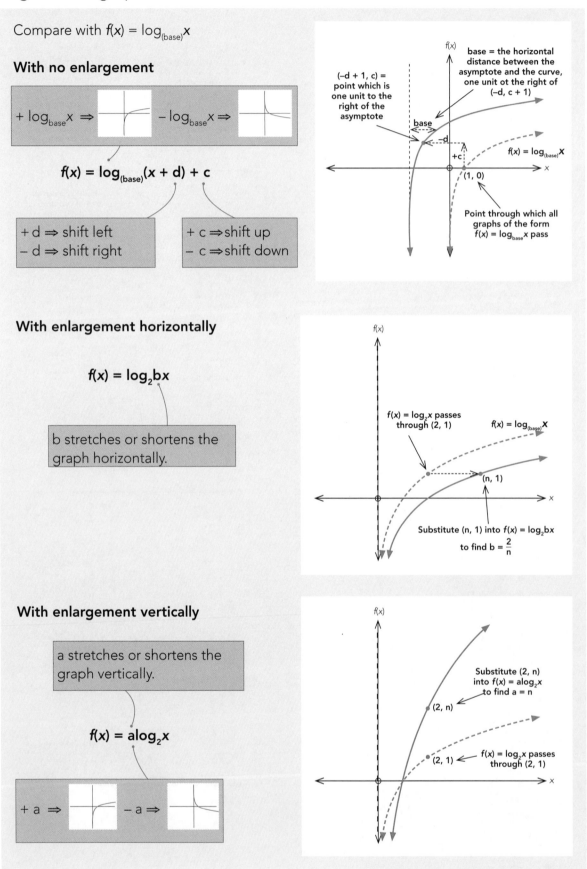

Logarithmic graphs

Compare with $f(x) = \log_{(base)}x$

With no enlargement

$+ \log_{base}x \Rightarrow$ | $- \log_{base}x \Rightarrow$

$$f(x) = \log_{(base)}(x + d) + c$$

$+ d \Rightarrow$ shift left
$- d \Rightarrow$ shift right

$+ c \Rightarrow$ shift up
$- c \Rightarrow$ shift down

$(-d + 1, c) =$ point which is one unit to the right of the asymptote

base = the horizontal distance between the asymptote and the curve, one unit ot the right of $(-d, c + 1)$

base

$-d$

$+c$

$f(x) = \log_{(base)}x$

$(1, 0)$

Point through which all graphs of the form $f(x) = \log_{base}x$ pass

With enlargement horizontally

$$f(x) = \log_2 bx$$

b stretches or shortens the graph horizontally.

$f(x) = \log_2 x$ passes through $(2, 1)$

$f(x) = \log_{(base)}x$

$(n, 1)$

Substitute $(n, 1)$ into $f(x) = \log_2 bx$ to find $b = \dfrac{2}{n}$

With enlargement vertically

a stretches or shortens the graph vertically.

$$f(x) = a\log_2 x$$

$+ a \Rightarrow$ | $- a \Rightarrow$

Substitute $(2, n)$ into $f(x) = a\log_2 x$ to find $a = n$

$(2, n)$

$(2, 1)$

$f(x) = \log_2 x$ passes through $(2, 1)$

ISBN: 9780170415989

Square root graphs

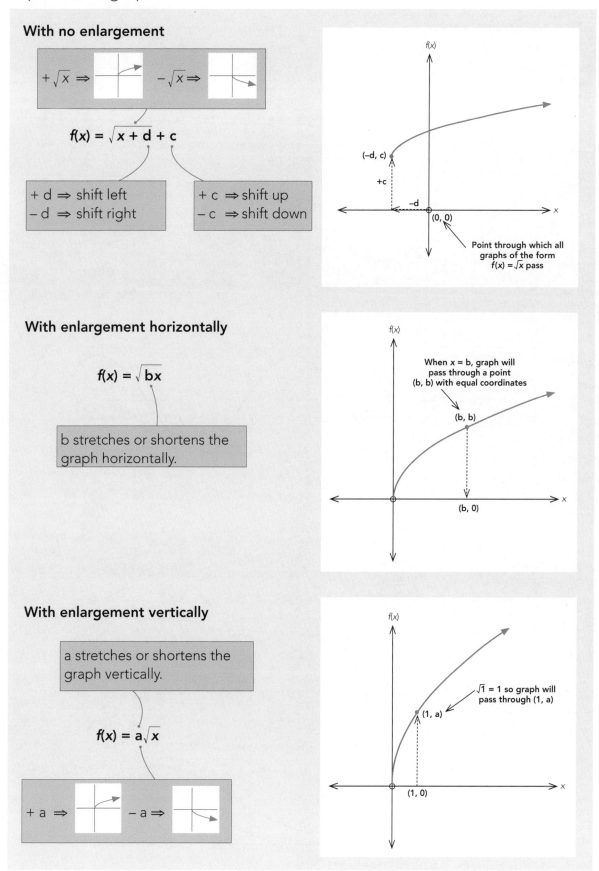

With no enlargement

$+\sqrt{x} \Rightarrow$

$-\sqrt{x} \Rightarrow$

$$f(x) = \sqrt{x + d} + c$$

$+ d \Rightarrow$ shift left
$- d \Rightarrow$ shift right

$+ c \Rightarrow$ shift up
$- c \Rightarrow$ shift down

$f(x)$

$(-d, c)$

$+c$

$-d$

$(0, 0)$

Point through which all
graphs of the form
$f(x) = \sqrt{x}$ pass

With enlargement horizontally

$$f(x) = \sqrt{bx}$$

b stretches or shortens the
graph horizontally.

$f(x)$

When $x = b$, graph will
pass through a point
(b, b) with equal coordinates

(b, b)

$(b, 0)$

With enlargement vertically

a stretches or shortens the
graph vertically.

$$f(x) = a\sqrt{x}$$

$+ a \Rightarrow$

$- a \Rightarrow$

$f(x)$

$\sqrt{1} = 1$ so graph will
pass through $(1, a)$

$(1, a)$

$(1, 0)$

Trigonometric graphs

Sine and cosine graphs
Compare with graphs of sin(*x*) or cos(*x*)

Note:
1 There are often several possible answers for these.
2 You do not usually need to consider reflections in the *x* or *y* axes, because sine and cosine graphs are symmetrical.

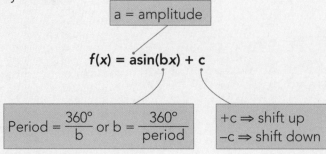

$$f(x) = a\sin(bx) + c$$

a = amplitude

$\text{Period} = \dfrac{360°}{b}$ or $b = \dfrac{360°}{\text{period}}$

+c ⟹ shift up
−c ⟹ shift down

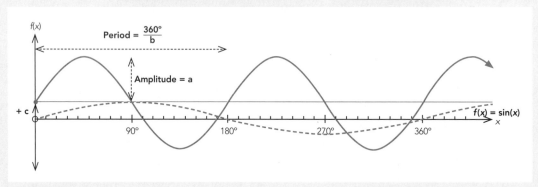

Cosine graphs

a = amplitude

+ a ⟹   − a ⟹

+ c ⟹ shift up
− c ⟹ shift down

$$f(x) = a\cos b(x - d) + c$$

$b = \dfrac{360°}{\text{period}}$

+ d ⟹ shift left
− d ⟹ shift right

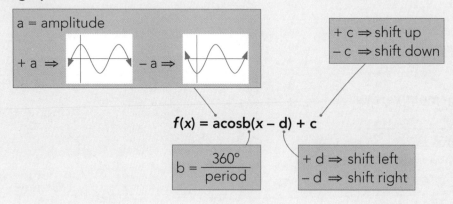

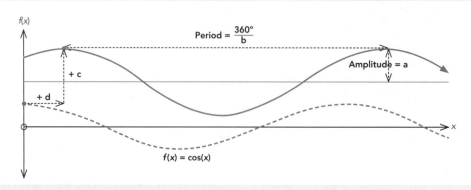

Sine graphs

a = amplitude

+a ⇒ −a ⇒

+ c ⇒ shift up
− c ⇒ shift down

$$f(x) = a\sin b(x - d) + c$$

$$b = \frac{360°}{\text{period}}$$

+ d ⇒ shift left
− d ⇒ shift right

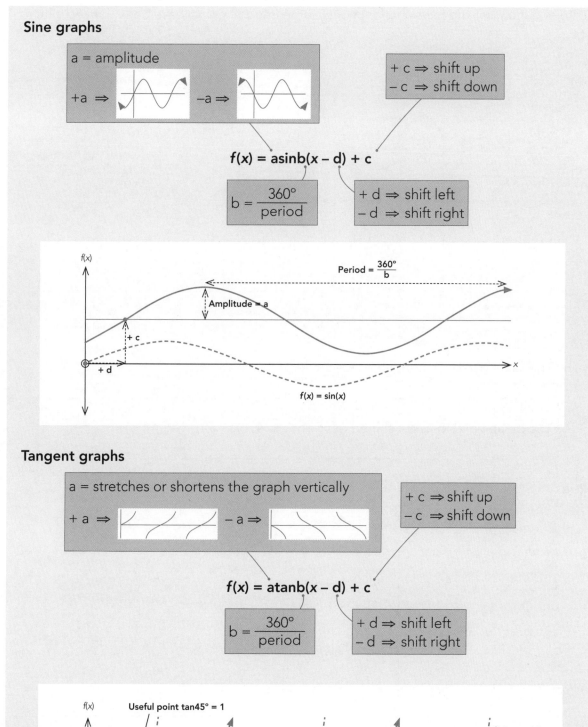

Period = $\frac{360°}{b}$

Amplitude = a

+ c

+ d

$f(x) = \sin(x)$

Tangent graphs

a = stretches or shortens the graph vertically

+ a ⇒ − a ⇒

+ c ⇒ shift up
− c ⇒ shift down

$$f(x) = a\tan b(x - d) + c$$

$$b = \frac{360°}{\text{period}}$$

+ d ⇒ shift left
− d ⇒ shift right

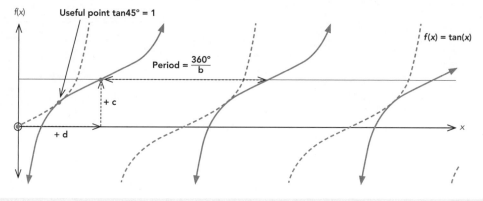

Useful point tan45° = 1

Period = $\frac{360°}{b}$

$f(x) = \tan(x)$

+ c

+ d

Write equations for the following functions. Some integral points have been marked with a dot.

1 a

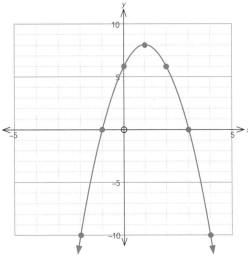

_____

b The same parabola after it has been reflected in the x-axis.

_____

c The parabola from **a**, but shifted left 5 and up 4.

_____

2 a

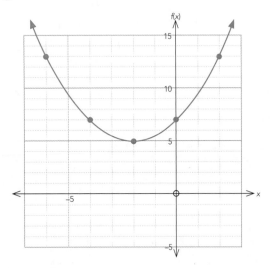

_____

b The same parabola after it has been reflected in the f(x) axis.

_____

c A parabola with the same minimum as the parabola in **a**, but with an f(x) intercept at (0, 6).

_____

3 a

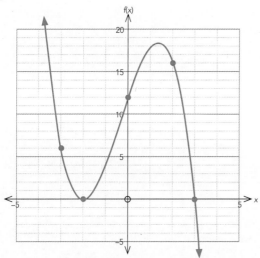

b The cubic with the same *x* intercepts, but which crosses the *f(x)* axis at (0, 6).

c The same cubic as **a**, but after it has been reflected in the line *f(x)* = 5. (Hint: Points on the *x*-axis will need to shift up 10.)

4 a

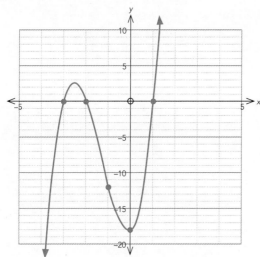

b The same cubic after it has been reflected in the *x*-axis.

c Reflect the cubic from **b** in the *y*-axis.

5 a

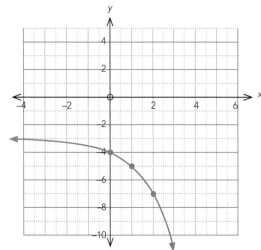

b Shift this graph 4 units to the right and up 1 unit.

_____

c Reflect your answer from **b** in the *x*-axis. (Hint: The asymptote will have to move from −c to +c.)

_____

6 a

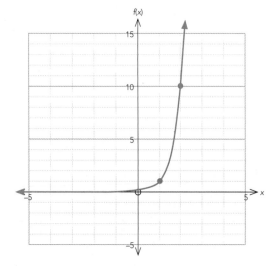

b Reflect the graph in the *f(x)* axis.

_____

c Stretch the graph in **b** so that each *f(x)* coordinate doubles.

_____

7 a

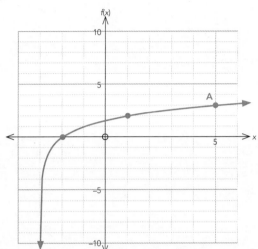

b Reflect the graph in the line $f(x) = 3$. (Hint: Think about where the point (−2, 0) will shift to.)

c Shift the graph in **a** so that the marked point A passes through the origin.

8 a

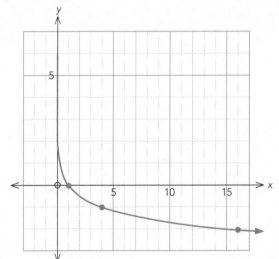

b Change the graph so that the x coordinate of each point is halved.

c For the graph you found in **b**, reflect it in the line $x = 2$.

9 a

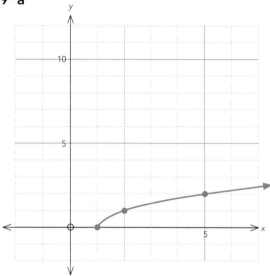

b Change the equation so that every **y** coordinate is tripled.

_____

c The same graph as in **b**, but shifted left 2 and up 5.

_____

10 a The absolute value graph which has a minimum at (−3, 5) and which passes through the point (3, 8).

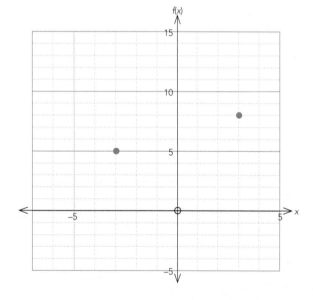

b Modify your answer to **a** so that the minimum point remains in the same place, but the gradient of each 'arm' is doubled.

_____

c Reflect the graph in **b** in the line $f(x) = 5$.

_____

ISBN: 9780170415989

11 a Base graph: $f(x) = \sin(x)$

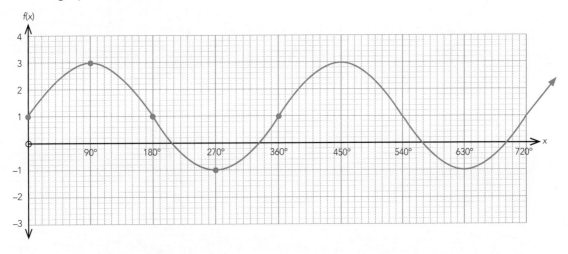

_____

b Modify the base graph equation so that the period is doubled.

_____

c Reflect your answer from **b** in the line $f(x) = 1$.

_____

12 a Base graph: $y = \cos(x)$

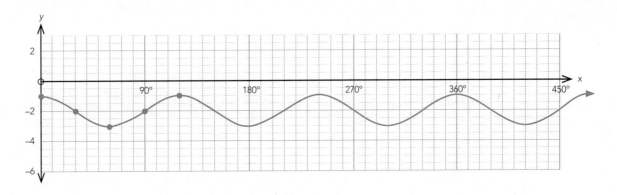

_____

b Shift the graph 90° to the right.

_____

c Make the amplitude equal to 4.

_____

13 a

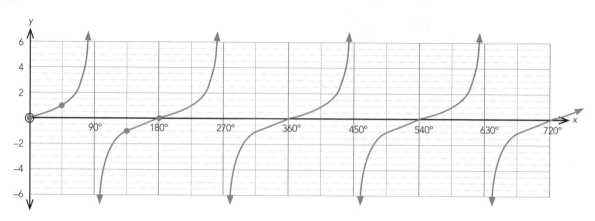

_____

b Shift the graph up 3.

_____

c Stretch the graph vertically by a factor of 2.

_____

14 a Base graph: $f(x) = \cos(x)$

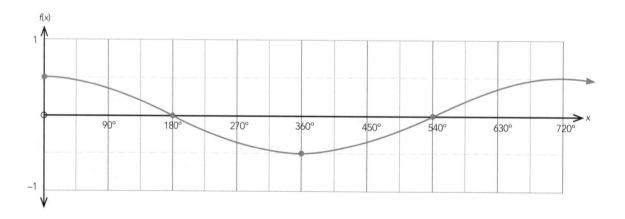

_____

b Reflect the graph in the line $f(x) = -1$.

_____

c Shift the graph from **b** 180° to the left.

_____

ISBN: 9780170415989

15a Base graph: $y = \sin(x)$

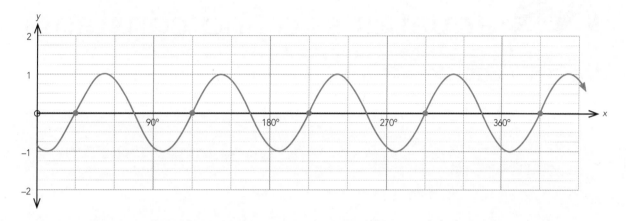

b Modify your answer so the amplitude is 3.

c Shift the graph in **a** up 2 units.

16a

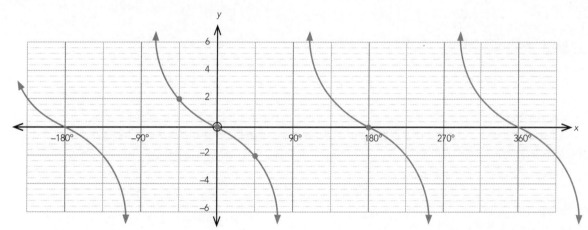

b Reflect the graph in the *y*-axis. What do you notice about the result?

c Shift the graph from **a** to the right by 60° and up 1 unit.

Using simultaneous equations to find constants

- You may need to use the coordinates of two points to help you find the values of constants in the equation for a function.

Example 1:

A function which takes the form $y = 0.1x^2(x - b) + c$ passes through the points (6, 4) and (10, 44). Find the values of b and c, and use them to write down the equation for the function.

Step 1: Substitute (6, 4) into $y = 0.1x^2(x - b) + c$:

$$4 = 0.1 \times 36 \times (6 - b) + c$$
$$4 = 21.6 - 3.6b + c$$
$$c = 3.6b - 17.6 \qquad \text{①}$$

> Hint: c does not have a coefficient, so writing both relationships as $c = ...$ is often an easy way to solve these.

Step 2: Substitute (10, 44) into $y = 0.1x^2(x - b) + c$:

$$44 = 0.1 \times 100 \times (10 - b) + c$$
$$44 = 100 - 10b + c$$
$$c = 10b - 56 \qquad \text{②}$$

Step 3: Solve ① and ②:

$$3.6b - 17.6 = 10b - 56$$
$$6.4b = 38.4$$
$$b = 6$$

Step 4: Substitute b = 6 into ②:

$$c = 10 \times 6 - 56$$
$$c = 4$$

Step 5: Write the completed equation: $\quad y = 0.1x^2(x - 6) + 4$

Example 2:

A function which takes the form $y = a(b^x)$ passes through the points (2, 8100) and (4, 6561). Find the values of a and b, and use them to write down the equation for the function.

Step 1: Substitute (2, 8100) into $y = a(b^x)$: $\qquad 8100 = a(b^2) \qquad \text{①}$

Step 2: Substitute (4, 6561) into $y = a(b^x)$: $\qquad 6561 = a(b^4) \qquad \text{②}$

Step 3: Divide ② by ①:

$$\frac{6561}{8100} = \frac{a(b^4)}{a(b^2)}$$

> Since there is just one term on each side of the = sign, and there is a common factor, **divide** one equation by the other.

$$0.81 = b^2$$
$$\therefore b = 0.9$$

Step 4: Substitute for b in ①: $\qquad 8100 = a(0.9^2)$

$$\therefore a = \frac{8100}{0.9^2} = 10\,000$$

Step 5: Write the completed equation: $\quad y = 10\,000(0.9)^x$

 ISBN: 9780170415989

Find equations for the following functions.

1 A function which takes the form $y = ax^2 + c$ passes through the points (10, 18) and (20, 48). Find the values of a and c, and use them to write down the equation for the function.

_____

_____

_____

_____

_____

_____

_____

_____

2 A function which takes the form $f(x) = a(b^x)$ passes through the points (3, 1) and (9, 64). Find the values of a and b, and use them to write down the equation for the function.

_____

_____

_____

_____

_____

_____

_____

_____

3 A function which takes the form $y = 0.05x^2(x - b) + c$ passes through the points (7, 6) and (10, 21). Find the values of b and c, and use them to write down the equation for the function.

_____

_____

_____

_____

_____

_____

_____

_____

_____

4 A function which takes the form $f(x) = a(r^{x-1})$ passes through the points (3, 490) and (6, 168.07). Find the values of a and r, and use them to write down the equation for the function.

5 A function which takes the form $y = \dfrac{a}{x-5} + c$ passes through the points (6, 16) and (11, 6). Find the values of a and c, and use them to write down the equation for the function.

6 A function which takes the form $y = a + b\sqrt{x}$ passes through the points (16, 102) and (36, 93). Find the values of a and b, and use them to write down the equation for the function.

 Applications

Complete the following statements and calculations, and answer the questions.

1   Charlie wants to model the rate of cooling of a cup of boiling milk. The milk boiled at a temperature of 101°C, and after 5 minutes the temperature was 73°C.

The temperature of the milk can be represented by $T = a(b)^t$

where T is temperature in °C
and t is the cooling time in minutes.

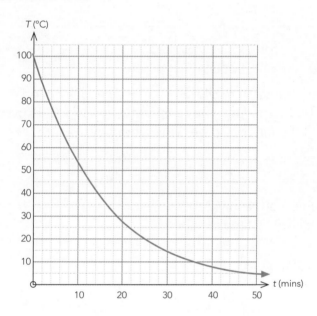

a The shape of this graph suggests that the value of b is greater/less than 1.

b The T intercept is _____, and it occurs when t = _____
This tells me that a = _____

c Charlie found that after 5 minutes the temperature was 73°C. Substitute these values into $T = a(b)^t$ to find the value of b.

_____

_____

_____ b = _____

d Write Charlie's model for the cooling of the milk: T = _____

e Using the model, calculate the temperature of the milk after one hour:

T = _____

= _____

f What would the temperature of the milk be after 24 hours?

_____

_____ Temperature = _____

g I think that this model is a suitable/an unsuitable model for this situation because

_____

_____

_____

h Describe how the graph would change for the cooling of alcohol, which has a boiling point of 78°C, and b = 0.86.

_____

_____

_____

Aroha obtained the same results as Charlie. The milk boiled at a temperature of 101°C, and after five minutes the temperature was 73°C. However, she thought the temperature of the milk could be represented by

$$T = \frac{840}{t + 10} + c$$

where T is temperature in °C
and t is the cooling time in minutes.

i Calculate the value of c.

_____

_____

_____

j Write Aroha's model for the cooling of the milk: $T =$ _____

k What would the temperature of the milk be after 24 hours?

_____

_____ Temperature = _____

l I think that this model is a suitable/an unsuitable model for this situation because

_____

_____

_____

2 Sam is analysing an image of a shark's fin. The peak of the shark's fin is at the point (16, 64).

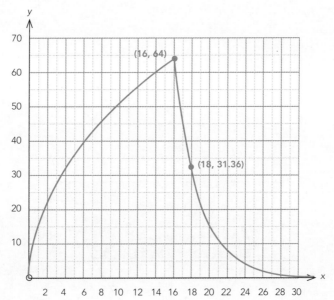

a She knows that for the front edge (between $x = 0$ and $x = 16$), the equation takes the form $y = \sqrt{ax}$. Calculate the value of a.

_____

_____

b Write the equation which models the front edge of the shark's fin.

_____

c The back edge of the fin is modelled by an equation which takes the form $y = pb^{(x-q)}$. Consider what happens when $x - q = 0$. Use this to work out the values of p and q.

_____

_____

_____

_____

_____

d She knows that the back edge also passes through the point (18, 31.36). Calculate the value of b.

_____

_____

_____

_____

e Write the equation which models the back edge of the shark's fin.

_____

3 Fred makes wooden bowls using a lathe. The sides of the bowl are parabolic in shape, but Fred cuts the bottom section off so that the base of the bowl is flat.
- The base of the bowl is 20 cm in diameter.
- The piece cut off is 10 cm deep.
- The bowl is 10 cm high.

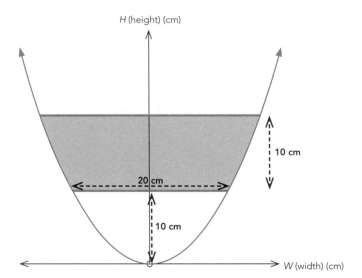

a Find the equation for the function which models the curved sides of the bowl.

_____

_____

_____

b Calculate the width of the top of the bowl.

_____

_____

Fred has a second but similar design for another bowl. Once again:
- the base of the bowl is 20 cm in diameter
- the piece cut off is 10 cm deep
- the bowl is 10 cm high.

For this bowl, the right side takes the shape of an exponential function of the form:

$$H = a(1.5)^w$$

c Calculate the value of a.

_____

_____

d Write down the equation for the function which models the curved right side of the bowl.

$$H = \underline{\hspace{2cm}}$$

e Assume that the bowl is symmetrical, so the left side is a mirror image of the right side. Write down the equation for the function which models the curved left side of the bowl.

$$H = \underline{\hspace{2cm}}$$

f Calculate the width of this bowl.

_____

_____

_____

Fred has a third similar design for yet another bowl. This time:
- the base of the bowl is 20 cm in diameter
- the piece cut off is **not** 10 cm deep
- the bowl is 10 cm high.

For this bowl, the right side takes the shape of a hyperbolic function of the form:

$$H = \frac{a}{W - 15} - 1$$

g Assuming the side of the bowl passes through the point (10, 10), calculate the value of a.

_____

_____

_____

h Write down the equation for the function which models the curved right side of the bowl.

$$H = \underline{\hspace{2cm}}$$

i Assume that the bowl is symmetrical, so the left side is a mirror image of the right side. Write down the equation for the function which models the curved *left* side of the bowl.

_____

_____

$$H = \underline{\hspace{2cm}}$$

j If the top of the bowl is formed by the line $f(x) = 20$, calculate the width of the top of this bowl.

_____

_____

k If the vertical height of the bowl is 10 cm, calculate how much Fred will need to cut off the bottom of his bowl.

_____

_____

_____

4   A student put some bacteria into a container of liquid food and placed it in a warm incubator for several days. She took very small samples at regular intervals and used the number of bacteria in each sample to estimate the total population in the container.

She knows that the equation for the population growth will take the form:

$$P = a \times b^t + c$$

where *P* is the number of bacteria in the population
and   *t* is the time in days.

The graph shows how the population grew:

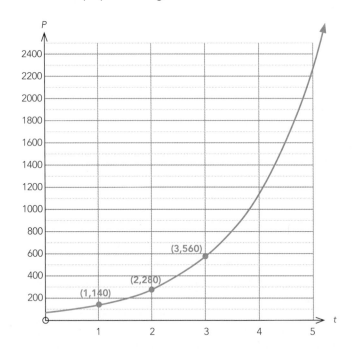

a Complete the table and use the ratios between each value for *P* to help you to find the value of b.

t	P
1	
2	
3	

b = _____

b Substitute the values for any two of the points into the equation $P = a \times b^t + c$ and solve the equations simultaneously in order to find the value of c.

_____

_____

_____ c = _____

c Find the value of a.

_____ a = _____

d Complete the equation for the population growth: P = _____

e What does the value of b tell you about the rate at which the population of bacteria is increasing?

_____

_____

f Use the graph to estimate the number of bacteria present after four days.

P ≈ _____

g Use the equation to calculate the size of the population after 4 days:

_____ P = _____

h How many bacteria were in the container at the start?

_____ P = _____

i Describe how the equation and the graph would change if all other conditions remained the same, but the population at the start was 200.

_____

_____

_____

_____

_____

_____

_____

j The student repeated the experiment, but for the first 24 hours she kept the container at a very low temperature, which prevented any growth of the population. Then she moved it to the warm incubator for the rest of the time. Assuming all other conditions remained the same, describe how the graph and the equation for the growth of the population would change.

_____

_____

_____

_____

_____

_____

_____

5 Huia owns a shop that sells cellphones. The new superphone is about to be released, and she has had 505 of these delivered. She will be unable to get more stock until at least 8 weeks after their release. She has been told that the equation for the number of superphones she has in stock is likely to take the form:

$$P = \frac{a}{8w + 2} + c$$

where P is the number of superphones she has in stock
and w is the number of weeks since their release.

The graph shows the number of superphones she has in stock for the first 7 weeks after their release.

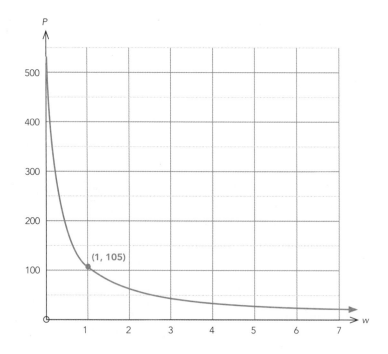

a Calculate the values of a and c.

_____

_____

_____

_____

_____

_____

_____

_____

b Complete the equation which models the decline in the number of superphones:

$P =$ _____

ISBN: 9780170415989

c How many superphones is she likely to have in stock at the start of week 6?

_____

_____

d Assume that she is unable to restock with superphones. According to this model, how many superphones would she have in stock after:

One year?

_____

_____

Five years?

_____

_____

e Do you feel this is a suitable model for the number of superphones that Huia has in stock? Refer to the equation in your explanation for your answer.

_____

_____

f In fact, Huia's sales of superphones do not follow this pattern, and she sells out of them completely at the end of week 5. She thinks that a quadratic model with a minimum at the end of week 5 would be appropriate for her sales. Find a suitable quadratic model for the sales of superphones.

_____

_____

_____

_____

g At the start of week 1 she has 232 superphones in stock. What does this suggest to you about the suitability of this quadratic model? Support your answer with calculations.

_____

_____

_____

h Compare these two models for the sales of superphones. You may wish to plot the quadratic model on the graph to help you.

_____

_____

_____

_____

_____

6 Toby has been left $10 000 by his grandmother. He invests it at an interest rate of x% per year compound interest.

He knows that the equation for the amount of money he has will take the form:

$$A = Pr^n$$

where A is the value ($) of his investment after n years
P is the value ($) of his investment at the start
r is the interest rate
and n is the number of years since the money was invested.

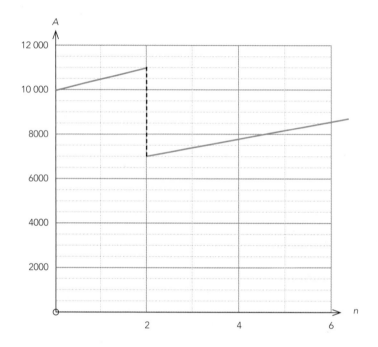

a The rate of interest remained constant over the first two years. His investment was worth $11 025 at the end of this period. Show that the rate of interest was 5% per year.

_____

_____

_____

b Write an equation which would enable him to calculate the value after n years of $10 000 invested at 5% per year compound interest.

_____

c If he left the money invested at this rate, what would its value be after 10 years?

_____

_____

d How long will it take for his investment to earn $1000?

_____

_____

e Write down the coordinates of the point at which his investment has doubled.

_____

_____

_____

_____

f Describe how the graph would differ if he had invested only $4000 at 5% per year.

_____

_____

_____

g What rate of interest would he need to get for his investment to double in value in 5 years?

_____

_____

_____

h After two years he withdrew $4000 to pay his polytech fees. He left the remaining money invested under the same conditions as during the first two years. Write an equation which would enable him to calculate the value of his remaining investment n years after he made his initial investment.

_____

_____

_____

_____

i Simple interest means that interest is paid only on the amount invested at the start, not on the total value of the investment. Write down the equation which would enable him to calculate the value of $10 000 invested at 5% simple interest for n years.

_____

_____

_____

j Describe similarities and differences between the graphs for $10 000 invested at 5% compound interest and 5% simple interest.

_____

_____

_____

_____

_____

7   Sefa has bought a new car for $40,000. His brother told him that its value over the first year he owns it can be modelled by an equation of the form:

$$V = ar^n$$

where V is the value of the car in dollars
        a is a constant
        r is the rate at which the car devalues each year
and n is the length of time he has owned it in years.

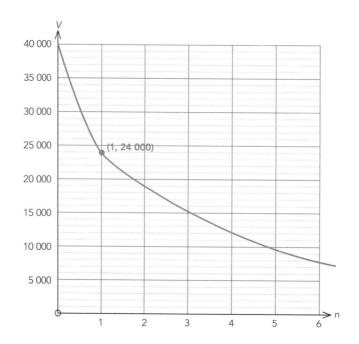

a   If the car is worth $24 000 after one year, calculate the values of a and r, and use them to complete the equation for the model for its value during the first year.

_____

_____

_____

_____

After one year the model changes to a form $V = ar^{0.5n}$
                where V is the value of the car in dollars
                        a is a different constant
                        r is a different rate
                and n is the length of time he has owned it in years.

b   If the car is worth $19 200 after 2 years, calculate the values of a and r, and use them to complete the equation for the model for its value after the first year.

_____

_____

_____

_____

A friend has told him that a better model for the value of his car is:

$$V = c - b\sqrt{n}$$

where V is the value of the car in dollars
       c is a constant
       b is a different rate
and n is the length of time he has owned it in years.

c According to this model, the car will have a value of $10 000 4 years after he buys it. Calculate the values of b and c, and use them to complete the equation for the model for the value of the car.

_____

_____

_____

_____

_____

_____

_____

d The car dealer told him that the value of the car would follow a quadratic model, and after 10 years the car would reach a minimum value of $10 000. Find the equation for this model.

_____

_____

_____

_____

_____

_____

_____

e Determine the value of the car 3 years after he bought it, using each of the three models.

_____

_____

_____

_____

_____

_____

_____

_____

8 Mei is designing a logo for her company which makes marbles. It is to take the shape of an 'M'. She has created three different versions for the right half of her proposed logo, which she plots for $0 \leq x \leq 18$.

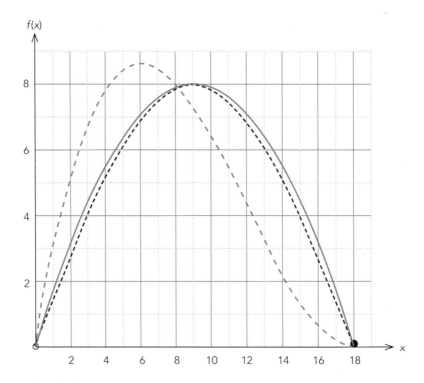

a The first version (——) of her design is formed by a quadratic function which
  - passes through the origin
  - passes through the point (18, 0) and
  - has a maximum point at (9, 8).

Write down the equation for this function.

_____

_____

_____

_____

_____

b The second version (------) of her design is formed by a cubic function which
  - passes through the origin
  - has a minimum point at (18, 0) and
  - passes through the point (8, 8).

Write down the equation for this function.

_____

_____

_____

_____

_____

c The third version (-------) of her design is formed by a trigonometric function which
- also passes through the origin
- has a maximum at (9, 8)
- passes through the point (18, 0) and
- is a sine function.

Write down the equation for this function.

_____

_____

_____

_____

_____

_____

d She wants to draw the left half of each logo for the domain $-18 \leq x \leq 0$. Write down the equation of each function after it has been reflected in the $f(x)$-axis.

   i   The quadratic function: _____

   ii The cubic function: _____

   iii The trigonometric function: _____

e Her finished logo (the completed 'M') will then be reflected in the x-axis. Write equations for the right side of the logo after it has been reflected in the x-axis.

   i   The quadratic function: _____

   ii The cubic function: _____

   iii The trigonometric function: _____

f Write equations for the left side of the logo after it has been reflected in the x-axis.

   i   The quadratic function: _____

   ii The cubic function: _____

   iii The trigonometric function: _____

9 Simon is designing a border. One unit of the border is shown in the graph. It has been constructed in three sections:

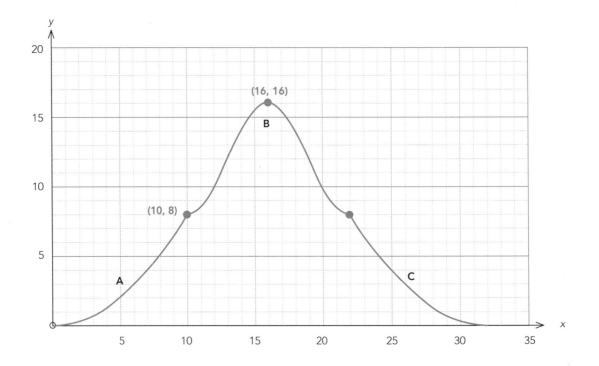

a Section A is a section of a parabola which has a minimum at the origin and which passes through the point (10, 8). Write down its equation.

_____

_____

_____

_____

b Section B is a section of a cosine curve which has a minimum at (10, 8) and a maximum at (16, 16). Write down its equation.

_____

_____

_____

_____

_____

_____

c Section C is the same as section A after it has been reflected in the line **x** = 16. Write down its equation.

_____

_____

_____

_____

 ISBN: 9780170415989

The basic unit is to be repeated to form the border. It is shown joined to two more units in the graph below. Write down the equation for each of the remaining sections.

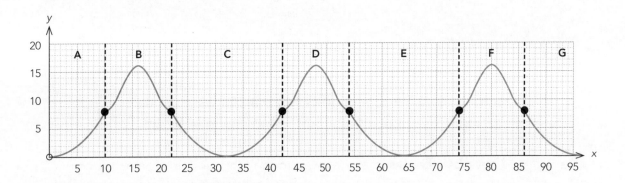

d Section D:

_____

_____

_____

_____

e Section E:

_____

_____

_____

_____

f Section F:

_____

_____

_____

_____

g Section G:

_____

_____

_____

_____

10 A fundraising campaign is launched to help families affected by a major flood. The rate at which donations are made can be modelled by the equation:

$$D = a\log_{10}(n + 2) + c$$

where D is the total value of the donations **in thousands of dollars**
     a is a constant
     c is a constant
and n is the number of days since the campaign started.

a After 18 days, $40 000 has been donated to the fund, and no money has been withdrawn. Assuming that there was nothing in the fund at the start, calculate the values of a and c, and use them to complete the equation for the amount of money in the fund.

_____

_____

_____

_____

b After 18 days, when $40 000 has been donated to the fund, there is some extremely bad publicity, and donations to the fund stop completely. Write an equation for the value of the fund from the eighteenth day onwards, asssuming that no money is withdrawn from it.

_____

c From day 20, the fund organisers start to distribute food each day to people worst affected by the flood, so that on day 21 there was $38 500 remaining. Assuming that they distribute food worth the same amount each day, and that there are no more donations to the fund, write down an equation which models the value of the fund from day 20 onwards.

_____

_____

d On what day will the money run out, and how much will they have to distribute on the final day?

_____

_____

e The organisers had hoped that the government would donate $5000 to kick-start the campaign. Assuming that this had happened, and the rate at which other donations were made was not affected, write an equation which would model the growth of the fund.

_____

_____

 Practice tasks

Practice task one

Part A

A new roller coaster ride is proposed for a theme park. In order to describe the ride to the investors, the equation of each section is needed so that an accurate diagram can be constructed. You need to:
- develop equations that model each section of the ride using the variables
  - H (height above the ground in m) and
  - d (distance from the start in m)
- describe the method you use to find each equation
- identify the domain for each equation.

Section 1: For the first 20 m the roller coaster follows a cubic model which:
- has a local minimum at the ground level at the start of the ride
- passes through the point (5, 5)
- reaches a peak and
- then descends to a height of 20 m above the ground.

Section 2: For the next 10 m it follows a quadratic model which:
- finishes at the model's lowest point at 10 m above the ground.

Section 3: Between 30 m and 45 m from the start, the roller coaster follows a trigonometric model which:
- takes the form $H = A \sin B(d - C) + D$
- is at its highest at the point (30, 10) and
- at its lowest when it is 36 m from the start, and at a height of 4 m.

Section 4: From the point (45, 7) the remainder of the ride is modelled by a hyperbolic function which:
- takes the form $H = \dfrac{A}{d - 39} + C$ and

- reaches the ground 59 m from the start of the ride.

Part B

The investors are not happy with Section 4 of the ride. They would like alternative models which have the same start and end points as in Part A. They suggest that the shape of this section of the ride could also be modelled by any of these functions: a parabola which has a minimum point at (59, 0), or an exponential function of the form $H = ab^d - 1$.
- Give the equation of a suitable function for each of these models.
- With respect to the roller coaster ride, discuss any relevant features or properties for each of the functions you have developed.

ISBN: 9780170415989

ISBN: 9780170415989

Practice task two

Part A

A new range of lampshades is being designed. When seen in cross-section, the new shades will be parabolic and hyperbolic.

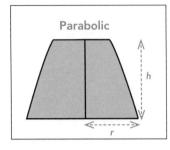
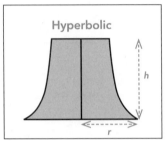

The equations of both sides of each shade are needed so that accurate diagrams can be constructed. All the shades have:
- a vertical height of 15 cm
- a base diameter of 30 cm and
- a top diameter of 10 cm.

You need to:
- develop equations that model the right and left sides of the lamp shades where:
  - h represents the vertical height of the lampshade in centimetres
  - r represents its radius in centimetres
  - the h-axis is the vertical line of symmetry of the shade and
  - the base of the shade lies along the r-axis
- describe the method you use to find each equation
- identify the domain for each equation.

Hints:
- The hyperbolae take the form $h = \dfrac{a}{r + b} + c$.
- The centre of rotation for the right side of the hyperbola is at (3, –3) and that for the left side is at (–3, –3).

Part B

It is decided that the height and the shapes of the sides are perfect, but the base diameter and the top diameter must both be reduced by 4 cm.
- Write new equations for both sides of each lampshade.
- Describe the transformation required in order to obtain the line and the hyperbola from their equivalents in Part A.
- Identify the domain for each new equation.

Part C

The designer would like to produce general equations for the *right* side only of each lampshade which he could use to produce lampshades that have the same sides as those in Part A, but with the top and bottom diameters both reduced by n cm.
- For each lampshade, write general equations (in terms of h, r and n) using the equations developed in Part A as a base.

 ISBN: 9780170415989

ISBN: 9780170415989

Practice task three

Part A

Anna is investigating laying up to 20 cubic metres of concrete around her new house. She is considering using two different companies. You need to:

- develop equations that model the concrete costs for each company, using the variables
  - C (cost of the concrete in $) and
  - v (volume of concrete in m³)
- describe the method you use to find each equation
- identify the domain for each equation
- recommend whether she should use Carl's Concrete or Connie's Concrete, depending on how much concrete she lays.

Carl's Concrete charges follow quadratic models.

- Up to and including 5 cubic metres (up to one truckload) he charges
  - at least $400
  - $700 for 5 cubic metres, which forms the model's highest point.
- For over 5 and up to and including 10 cubic metres (up to two truckloads) his charges follow the same model as the one you found for up to 5 cubic metres, but the model is
  - translated up $300 and
  - to the right by 5 cubic metres.
- For over 10 cubic metres (which requires bigger trucks) his charges
  - are at least $1100
  - follow a model which crosses the v-axis at 0 and 40.

Connie's Concrete charges follow hyperbolic models.

- Up to and including 5 cubic metres her charge
  - is at least $300
  - follows a model of the type $C = \dfrac{a}{v + b} + 100$
  - is $700 for 5 cubic metres.
- For over 5 cubic metres her charge
  - is at least $700
  - follows a model of the type $C = \dfrac{a}{v} + b$
  - is $1100 for 10 cubic metres.

Part B

Kauri would like to set up a company to compete with Carl and Connie. He thinks that he could charge less than the others for quantities up to 20 cubic metres.

- He would like to model his charges on a sine curve.
- The model should have a minimum value where $v = 0$ and a maximum value where $v = 20$.
- The least he can afford to charge for a load is $200.

Find an equation to model this situation, and include ranges of values for any constants which can be varied. Justify your choice of values for the constants.

ISBN: 9780170415989

Answers

Functions (pp. 6–14)

1 Function: Yes
Domain: Real numbers
Range: Real numbers
2 Function: Yes
Domain: Real numbers
Range: $f(x) \leq 5$
3 Function: Yes
Domain: Real numbers except for –3
Range: Real numbers except for 1
4 Function: No
5 Function: Yes
Domain: Real numbers
Range: $f(x) > 0$

Calculations using function notation (p. 9)

1 $f(2) = 2$
$f(-4) = -16$
$f(0) = -4$
2 $f(2) = 9$
$f(-4) = -3$
$f(0) = -3$
3 $f(2) = 4$
$f(-4) = \dfrac{1}{16}$
$f(0) = 1$
4 $x = 12.5$
5 $x = 5$ or –2

Features of functions (pp. 10–15)

1 Maximum: None, but local maximum (1, 4)
Minimum: None, but local minimum (–1, 0)
x intercept: (–1, 0) and (2, 0)
y intercept: (0, 2)
Asymptotes: None
Axes of symmetry: None
Centre of rotational symmetry: (0, 2)
Domain: Real numbers
Range: Real numbers
2 Maximum: None
Minimum: None
x intercept: (6, 0)
y intercept: None
Asymptotes: Vertical: $x = 0$; Horizontal: $y = 1$
Axes of symmetry: $y = x + 1$ and $y = -x$
Centre of rotational symmetry: (0, 1)
Domain: Real numbers except 0
Range: Real numbers except 1

3 Maximum: None
Minimum: slighty above 2
x intercept: None
y intercept: (0, 3)
Asymptotes: $y = 2$
Axes of symmetry: None
Centre of rotational symmetry: None
Domain: Real numbers
Range: $y > 2$
4 Maximum: 3
Minimum: –3
x intercept: …, (–180°, 0), (–90°, 0), (0°, 0), (90°, 0), (180°, 0), …
y intercept: (0, 0)
Asymptotes: None
Axes of symmetry: $x =$ …, –135°, –45°, 135°, 45°, …
Centre of rotational symmetry: $x =$ …, (–180°, 0), (–90°, 0), (0°, 0), (90°, 0), (180°, 0), …
Domain: Real numbers
Range: $-3 \leq y \leq 3$
Amplitude: 3
Period: 180°

Parabolas: revision (pp. 16–25)

1

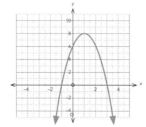

2

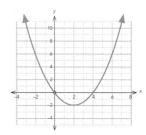

3

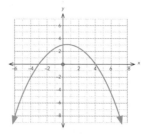

4 $y = -(x - 2)^2 - 1$ **5** $f(x) = \dfrac{1}{2}(x + 1)(x + 6)$

6 $y = -\dfrac{1}{4}(x + 2)^2 - 3$ **7** $f(x) = 0.1(x + 2)(x - 5)$

8 $y = -0.2(x + 6)(x - 2)$ **9** $y = 0.05(x - 6)^2 - 3$

ISBN: 9780170415989

Plotting graphs (pp. 26–44)

1 Cubic graphs (pp. 26–28)

a

x	$f(x) = x^2(x - 4)$
5	25
4	0
3	–9
2	–8
1	–3
0	0
–1	–5
–2	–24
–3	(–63)

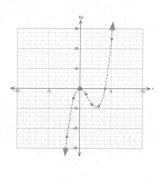

b

x	$f(x) = (x + 1)(x - 1)(x - 3)$
4	15
3	0
2	–3
1	0
0	3
–1	0
–2	–15
–3	(–48)

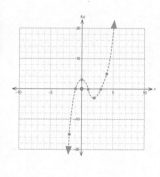

2 Quartic graphs (pp. 29–31)

a

x	$f(x) = x^2(x - 2)(x + 3)$
3	(54)
2	0
1	–4
0	0
–1	–6
–2	–16
–3	0

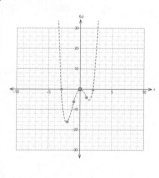

b

x	$f(x) = (x + 1)(x + 2)(x - 1)(x - 3)$
3	0
2	–12
1	0
0	6
–1	0
–2	0
–3	48

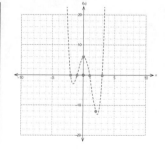

3 Rectangular hyperbolae (pp. 32–33)

a

x	$f(x) = \dfrac{6}{x}$
3	2
2	3
1	6
0	Undefined
–1	–6
–2	–3
–3	–2
6	1
–6	–1

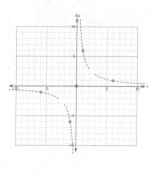

b

x	$f(x) = \dfrac{10}{x}$
10	1
5	2
2	5
1	10
0	Undefined
–1	–10
–2	–5
–5	–2
–10	–1

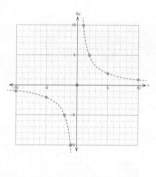

4 Exponential graphs (pp. 34–37)

a   i

x	$f(x) = 3^x$
3	27
2	9
1	3
0	1
–1	$\dfrac{1}{3}$
–2	$\dfrac{1}{9}$
–3	$\dfrac{1}{27}$

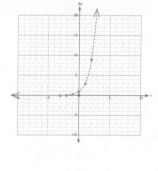

ii

x	$f(x) = 10^x$
3	1000
2	100
1	10
0	1
–1	0.1 or $\dfrac{1}{10}$
–2	0.01 or $\dfrac{1}{100}$
–3	0.001 or $\dfrac{1}{1000}$

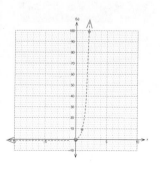

b i

x	$f(x) = \left(\frac{1}{4}\right)^x$
3	$\frac{1}{64}$
2	$\frac{1}{16}$
1	$\frac{1}{4}$
0	1
−1	4
−2	16
−3	64

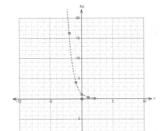

ii

x	$f(x) = \left(\frac{1}{10}\right)^x$
3	$0.001 \text{ or } \frac{1}{1000}$
2	$0.01 \text{ or } \frac{1}{100}$
1	$0.1 \text{ or } \frac{1}{10}$
0	1
−1	10
−2	100
−3	1000

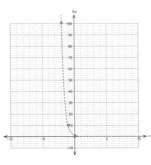

5 Logarithmic graphs (pp. 38–39)

a

x	$y = \log_3 x$
9	$\log_3(9) = \log_3(3^2) = 2$
3	$\log_3(3) = \log_3(3^1) = 1$
1	$\log_3(1) = \log_3(3^0) = 0$
0	$\log_3(0)$ is undefined
$\frac{1}{3}$	$\log_3\left(\frac{1}{3}\right) = \log_3(3^{-1}) = -1$
$\frac{1}{9}$	$\log_3\left(\frac{1}{9}\right) = \log_3(3^{-2}) = -2$

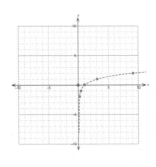

b

x	$y = \log_{10} x$
100	$\log_{10}(100) = \log_{10}(10^2) = 2$
10	$\log_{10}(10) = \log_{10}(10^1) = 1$
1	$\log_{10}(1) = \log_{10}(10^0) = 0$
0	$\log_{10}(0)$ is undefined
$\frac{1}{10}$	$\log_{10}\left(\frac{1}{10}\right) = \log_{10}(10^{-1}) = -1$
$\frac{1}{100}$	$\log_{10}\left(\frac{1}{100}\right) = \log_{10}(10^{-2}) = -2$

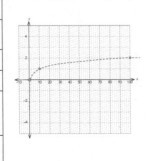

8 Trigonometric graphs (pp. 42–44)

Graph of $f(x) = \sin x$

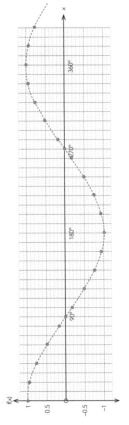

Graph of $f(x) = \cos x$

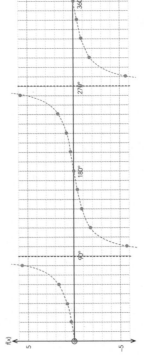

Graph of $f(x) = \tan x$

   ISBN: 9780170415989

Transformations of graphs (pp. 46–72)

Translations (pp. 46–54)

1 Vertical translations (pp. 46–48)

1

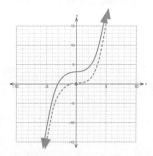

Domain: Real numbers
Range: Real numbers
Transformation from $f(x) = x^3$: Translation up by 3 units

2

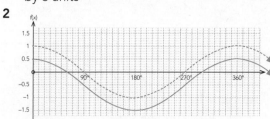

Domain: Real numbers
Range: $-1.5 \leq f(x) \leq 0.5$
Transformation from $f(x) = \cos(x)$: Translation down by 0.5 units

3

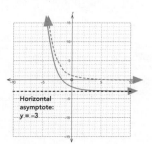

Domain: Real numbers
Range: $f(x) > -3$
Transformation from $f(x) = (\frac{1}{2})^x$: Translation down by 3 units

4

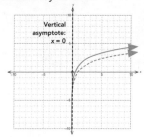

Domain: $x > 0$
Range: Real numbers
Transformation from $y = \log_2 x$: Translation up by 1 unit

2 Horizontal translations (pp. 49–51)

1

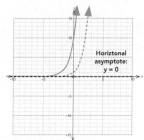

Domain: Real numbers
Range: $y > 0$
Transformation from $y = 3^x$: Translation left by 2 units

2

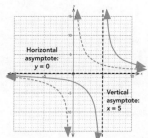

Domain: Real numbers except 5
Range: Real numbers except 0
Transformation from $xy = 12$: Translation right by 5 units

3

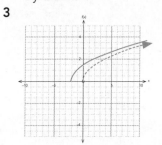

Domain: $x \geq -2$
Range: $f(x) = \geq 0$
Transformation from $f(x) = \sqrt{x}$: Translation left by 2 units

4

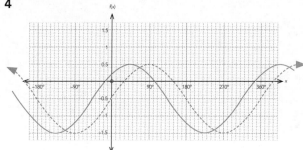

Domain: Real numbers
Range: $-1 \leq f(x) \leq 1$
Transformation from $f(x) = \sin(x)$: Translation right by 45°

3 Combinations of vertical and horizontal translations (pp. 52–54)

1

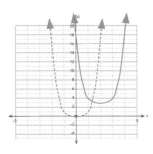

Domain: real numbers
Range: $f(x) \geq 3$
Transformation from $f(x) = x^4$: Translation to the right by 2 and up by 3.

2

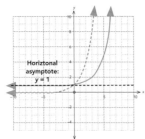

Domain: Real numbers
Range: $y > 1$
Transformation from $y = 2^x$: Translation to the right by 3 and up by 1.

3

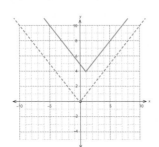

Domain: real numbers
Range: $f(x) \geq 4$
Transformation from $f(x) = |x|$: Translation to the right by 1 and up by 4

4

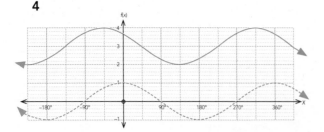

Domain: Real numbers
Range: $2 \leq f(x) \leq 4$
Transformation from $\cos(x)$: Translation to the left by 45°, and up by 3.

Mix and match (pp. 55–57)

1 A   **2** D
3 B   **4** C
5 D   **6** A
7 B   **8** C
9 D   **10** A or B
11 E   **12** B or A
13 C
14 $y = (x + 1)^3 - 2$ **15** $y = \sqrt{x - 1} + 3$
16 $y = |x - 4| + 3$ **17** $y = 3^{(x - 3)} + 5$
18 $f(x) = \cos(x - 45°) + 0.5$
19 $f(x) = \tan(x + 90°) - 1$ or $f(x) = \tan(x - 90°) - 1$

Reflections (pp. 58–62)
1 Reflections in the x-axis (pp. 58–60)

1

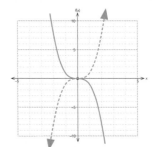

Domain: Real numbers
Range: Real numbers
Transformation from $f(x) = x^3$: Reflection in the x-axis.

2

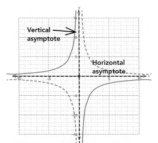

Domain: Real numbers except 0
Range: Real numbers except 0
Transformation from $y = \dfrac{8}{x}$: Reflection in the x (or y) axis.

3

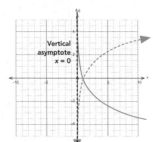

Domain: $x > 0$
Range: Real numbers
Transformation from $f(x) = \log_2 x$:
Reflection in the x-axis.

4

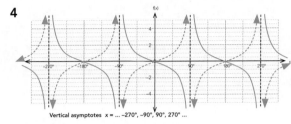

Vertical asymptotes x = ... –270°, –90°, 90°, 270° ...

Domain: Real numbers except …, –270, –90, 90, 270, …

Range: Real numbers

Transformation from $f(x) = \tan(x)$: Reflection in the x-axis.

2 Reflections in the y-axis (or f(x) axis) (pp. 61–62)

1

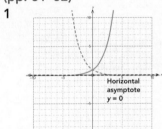

Horizontal asymptote y = 0

Domain: Real numbers

Range: $y > 0$

Transformation from $y = \left(\frac{1}{2}\right)^x$: Reflection in the y-axis.

2

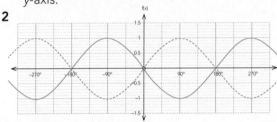

Domain: Real numbers

Range: $-1 \le f(x) \le 1$

Transformation from $f(x) = \sin(x)$: Reflection in the f(x) axis.

Combinations of translations and reflections (pp. 63–65)

1	B	**2**	C
3	A	**4**	D
5	C	**6**	D
7	B	**8**	A
9	A	**10**	C
11	D	**12**	B
13	B	**14**	D
15	A	**16**	C
17	$y = 5^{(-x)} - 4$	**18**	$y = \sqrt{-x} + 2$

19 $f(x) = -\tan(x + 45°) + 2$
or $f(x) = \tan(x - 135°) + 2$
or $f(x) = \tan(45° - x) + 2$
or $f(x) = \tan(-x - 135°) + 2$

20 $f(x) = -\cos(x + 45°) - 1$
or $f(x) = \cos(x - 135°) - 1$
or $f(x) = \cos(135° - x) - 1$

21 $\sin(x + 90°) + 0.5$ or $-\sin(x - 90°) + 0.5$
or $\sin(-x - 90°) + 0.5$

More cubic and quartic graphs (pp. 68–71)

1	B	**2**	C
3	D	**4**	A
5	A	**6**	D
7	B	**8**	C

9 $y = x^2(x - 3)$ or $y = -x^2(3 - x)$

10 $y = -x^2(x + 2)$

11 $y = -x(x - 2)(x + 1)$ or $y = x(2 - x)(x + 1)$

12 $y = x(x - 2)(x - 3)$ or $y = -x(2 - x)(x - 3)$
or $y = -x(x - 2)(3 - x)$

13 $y = (x + 1)(x - 2)(x - 3)$
or $y = -(x + 1)(2 - x)(x - 3)$
or $y = (x + 1)(x - 2)(3 - x)$

14 $y = (x - 1)(x + 2)(x - 3)(x + 4)$
or $y = -(1 - x)(x + 2)(x - 3)(x + 4)$
or $y = -(x - 1)(x + 2)(3 - x)(x + 4)$

15 $y = -(x - 1)(x + 1)(x + 2)(x - 4)$
or $y = (1 - x)(x + 1)(x + 2)(x - 4)$
or $y = (x - 1)(x + 1)(x + 2)(4 - x)$

16 $y = -(x - 1)^2(x + 2)(x + 3)$

Enlargements (pp. 70–77)
1 Vertical enlargements (pp. 70–72)

1

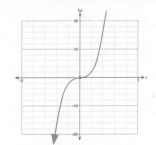

Domain: Real numbers

Range: Real numbers

Transformation from $f(x) = x^3$: Enlargement (stretch) along the f(x) axis

2

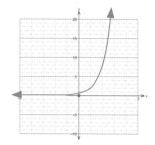

Domain: Real numbers
Range: $y > 0$
Transformation from $y = 4^x$: Enlargement (shortening) along the y-axis

3

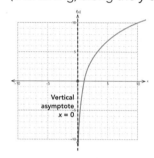

Domain: $x > 0$
Range: Real numbers
Transformation from $f(x) = \log_2 x$: Enlargement (stretch) along the $f(x)$ axis.

4

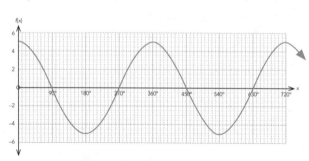

Domain: Real numbers
Period: 360°
Amplitude: 5
Range: $-5 \leq x \leq 5$
Transformation from $f(x) = \cos(x)$:
Enlargement (stretch) along the $f(x)$ axis.

2 Horizontal enlargements (pp. 73–75)

1

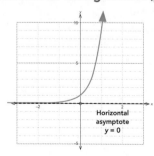

Domain: Real numbers
Range: $y > 0$
Transformation from $y = 3^x$: Enlargement (shortening) along the x-axis.

2

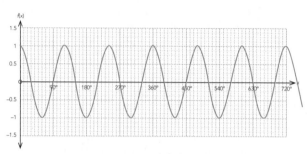

Domain: Real numbers
Period: 120°
Amplitude: 1
Range: $-1 \leq f(x) \leq 1$
Transformation from $f(x) = \cos(x)$:
Enlargement (shortening) along the x-axis.

3

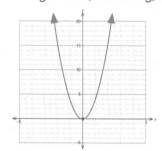

Domain: Real numbers
Range: $y \geq 0$
Transformation from $y = x^2$: Enlargement (shortening) along the x-axis.

4

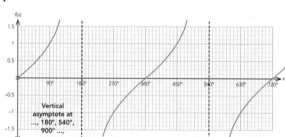

Domain: Real numbers except …, −180, 180, 540, …
Period: 360°
Amplitude: No fixed amplitude
Range: Real numbers
Transformation from $f(x) = \tan(x)$:
Enlargement (stretch) along the x-axis.

3 Combinations of vertical and horizontal enlargements (pp. 76–77)

1	C	2	B
3	D	4	A
5	A	6	D
7	C	8	B
9	D	10	A
11	E	12	B
13	C		

Putting it all together (pp. 78–93)

1 a $y = -2(x + 1)(x - 3)$
  b $y = 2(x + 1)(x - 3)$
  c $y = -2(x + 6)(x + 2) + 4$

2 a $f(x) = 0.5(x + 2)^2 + 5$
  b $f(x) = 0.5(2 - x)^2 + 5$
  c $f(x) = 0.25(x + 2)^2 + 5$

3 a $f(x) = -(x + 2)^2(x - 3)$
  b $f(x) = -0.5(x + 2)^2(x - 3)$
  c $f(x) = (x + 2)^2(x - 3) + 10$

4 a $y = 3(x + 2)(x - 1)(x + 3)$
  b $y = -3(x + 2)(x - 1)(x + 3)$
  c $y = -3(x - 2)(x + 1)(x - 3)$

5 a $y = -2^x - 3$
  b $y = -2^{(x - 4)} - 2$
  c $y = 2^{(x - 4)} + 2$

6 a $f(x) = 10^{(x - 1)}$
  b $f(x) = 10^{(-x - 1)}$
  c $f(x) = 2(10^{(-x - 1)})$

7 a $y = \log_2(x + 3)$
  b $y = -\log_2(x + 3) + 6$
  c $y = \log_2(x + 8) - 3$

8 a $y = -0.5\log_2 x$
  b $y = -0.5\log_2(2x)$
  c $y = -0.5\log_2(-2(x - 4))$

9 a $y = \sqrt{x - 1}$
  b $y = 3\sqrt{x - 1}$
  c $y = 3\sqrt{x + 1} + 5$

10 a $f(x) = 0.5|x + 3| + 5$
  b $f(x) = 0.5|2(x + 3)| + 5$
  c $f(x) = -0.5|2(x + 3)| + 5$

11 a $f(x) = 2\sin(x) + 1$
  b $f(x) = 2\sin(0.5x) + 1$
  c $f(x) = -2\sin(0.5x) + 1$

12 a $y = \cos(3(x)) - 2$
  b $y = \cos(3(x - 90°)) - 2$
  c $y = 4\cos(3(x - 90°)) - 2$

13 a $y = \tan(x)$
  b $y = \tan(x) + 3$
  c $y = 2\tan(x) + 3$

14 a $f(x) = 0.5\cos(0.5x)$
  b $f(x) = -0.5\cos(0.5x) - 2$
  c $f(x) = -0.5\cos(0.5(x + 180°)) - 1.5$

15 a $y = \sin(4(x - 30°))$
  b $y = 3\sin(4(x - 30°))$
  c $y = 3\sin(4(x - 30°)) + 2$

16 a $y = -2\tan(x)$
  b $y = -2\tan(-x)$
    This is the same as the graph of
    $y = 2\tan(x)$.
  c $y = -2\tan(-(x - 60°)) + 1$

Using simultaneous equations to find constants (pp. 94–96)

1 $(10, 18) \Rightarrow c = 18 - 100a$
  $(20, 48) \Rightarrow c = 48 - 400a$
  $\therefore a = 0.1$ and $c = 8$
  Function: $y = 0.1x^2 + 8$

2 $(3, 1) \Rightarrow 1 = a \times b^3$
  $(9, 64) \Rightarrow 64 = a \times b^9$
  $\therefore b = 2$ and $a = 0.125$
  Function: $f(x) = 0.125(2^x)$

3 $(7, 6) \Rightarrow c = 2.45b - 11.15$
  $(10, 21) \Rightarrow c = 5b - 29$
  $\therefore b = 7$ and $c = 6$
  Function: $y = 0.05x^2(x - 7) + 6$

4 $(3, 490) \Rightarrow 490 = a \times r^2$
  $(6, 168.07) \Rightarrow 168.07 = a \times r^5$
  $\therefore r = 0.7$ and $a = 1000$
  Function: $f(x) = 1000(0.7^{x - 1})$

5 $(6, 16) \Rightarrow a = 36 - 6c$
  $(11, 6) \Rightarrow a = 16 - c$
  $\therefore c = 4$ and $a = 12$
  Function: $y = \dfrac{12}{x - 5} + 4$

6 $(16, 102) \Rightarrow a = 102 - 4b$
  $(36, 93) \Rightarrow a = 93 - 6b$
  $\therefore b = -4.5$ and $a = 120$
  Function: $y = 120 - 4.5\sqrt{x}$

Applications (pp. 97–114)

1 a The shape of this graph suggests that the value of b is **less** than 1.
  b The T intercept is **101°**, and it occurs when $t = $ **0°**.
    This tells me that a = **101**.
  c $b = 0.9371$
  d $T = 101(0.9371)^t$
  e $T = 2.05°$
  f Temperature = 2.4×10^{-39} so it is effectively 0°C.

g I think that this model is **an unsuitable** model for this situation because the boiled milk will be very close to the temperature of its surroundings after 24 hours, and that is very unlikely to be 0°C.

h The graph would cut the T-axis at a lower point (78°), and it would drop more rapidly because the value for b is lower (0.86).

i c = 17

j $T = \dfrac{840}{t + 10} + 17$

k Temperature = 17.58°C.

l I think that this model is **a suitable** model for this situation because the boiled milk will be very close to the temperature of its surroundings after 24 hours, and that is very likely to be about 17° C.

2 a a = 256

b $y = \sqrt{256x}$

c $x - q = 0 \Rightarrow y = p \times b^0$, so $y = p$
Passes through (16, 64) $\Rightarrow 64 = pb^{(16-16)}$
so p = 64 and q = 16

d b = 0.7

e $y = 64(0.7^{x-16})$

3 a $H = 0.1W^2$

b Width of bowl = = 28.28 cm

c a = 0.1734

d $H = 0.1734(1.5)^w$

e $H = 0.1734(1.5)^{-w}$

f Width = $2 \times \dfrac{\log(20 \div 0.1734)}{\log 1.5} = 2 \times 11.71 = 23.42$ cm

g a = –55

h $H = \dfrac{-55}{W - 15} - 1$

i $H = \dfrac{-55}{15 - w} - 1$

j $2 \times 12.38 = 24.76$

k H intercept = (0, 2.7) so amount cut off = 7.3 cm

4 a

t	P
1	140
2	280
3	560

b = 2

b c = 0

c a = 70

d $P = 70(2^t)$

e The population of bacteria is doubling every day.

f $P \approx 1100$

g $P = 1120$

h $P = 70$

i The equation would become $P = 200(2^t)$
The intercept on the P-axis would increase to 200, and the graph would go up more steeply.

j Between 0 and 1, the graph would be horizontal, with the equation $P = 70$.
The curve on the original graph would shift 1 day to the right.
The equation would become $P = 70(2^{t-1})$

5 a a = 1000
c = 5

b $P = \dfrac{1000}{8w + 2} + 5$

c 25 superphones

d After one year: about 7 superphones
After five years: about 5 superphones

e This is not a suitable model for the number of superphones in stock because when w becomes very large, $\dfrac{1000}{8w + 2}$ becomes very small but never negative, so there will always be at least five superphones left unsold.

f Passes through (0, 505) and (5, 0) $\Rightarrow$
$P = 20.2(w - 5)^2$

g $w = 1 \Rightarrow P = 20.2(1 - 5)^2 = 232.2$
$\therefore$ This value suggests that the quadratic model is a very good one.

h For the hyperbolic model: her stock of superphones decreases very rapidly over the first two weeks, but then it flattens out, and it never drops below 5.
For the quadratic model: her stock drops much more slowly, over the first two weeks, but it drops much faster than the hyperbolic model after that, until there are none left at the end of the fifth week. After week 5, this model would not fit at all, because the stock of superphones would increase.

6 a $A = Pr^n$
$11\,025 = 10\,000 \times r^2$
$r^2 = 1.1205$
$\therefore r = 1.05$, so the interest rate is 5%.

b $A = 10\,000 \times 1.05^n$

c $16\,288.95

d 1.953 years

e (14.21, 20 000)

f The A intercept would be at 4000 rather than 10 000, and it would increase more slowly because less interest is added each year.

g 14.87%

h $A = 7\,025 \times 1.05^{n-2}$

i $A = 10\,000 + 500n$

j Compound interest graph: A intercept at (0, 10 000), and the gradient would keep increasing as n increases.
Simple interest graph: Also has an A intercept at (0, 10 000), but the gradient would stay at 500, however big n became.

7 a $(0, 40\,000) \Rightarrow a = 40\,000$
$(1, 24\,000) \Rightarrow r = 0.6$
$\therefore V = 40\,000(0.6)^n$

b $(1, 24\,000) \Rightarrow V = a \times r^{0.5}$
$(2, 19\,200) \Rightarrow V = a \times r^1$
so $r^{0.5} = 0.8, \Rightarrow r = 0.64$
$(1, 24\,000) \Rightarrow a = 30\,000$
$\therefore V = 30\,000(0.64)^{0.5n}$

c $(0, 40\,000) \Rightarrow c = 40\,000$
$(4, 10\,000) \Rightarrow b = 15\,000$
$\therefore V = 40\,000 - 15\,000\sqrt{n}$

d Equation is of the form $V = a(n - b)^2 + c$
Minimum at (10, 10 000) $\Rightarrow V = a(n - 10)^2 + 10\,000$
$(0, 40\,000) \Rightarrow a = 300$
$\therefore V = 300(n - 10)^2 + 1000$

e Brother: $15 360.00
Friend: $14 019.24
Car dealer: $24 700.00

8 a $(0, 0) \Rightarrow c = -0.30103a$
$(18, 40) \Rightarrow a = 40$ and $c = 12.04$
$\therefore D = 40 \log_{10}(n + 2) - 12.04$

b $D = 40$

c $D = 40 - 1.5n$

d On day 47, when they will have only $1000 to distribute.

e $D = 40 \log_{10}(n + 2) - 7.04$

9 a $f(x) = -0.09877x(x - 18)$

b $f(x) = 0.01x(x - 18)^2$

c $f(x) = 8\sin(10x)$

d **i** $f(x) = -0.09877x(x + 18)$
ii $f(x) = -0.01x(x + 18)^2$
iii $f(x) = 8\sin(-10x)$

e **i** $f(x) = 0.09877x(x - 18)$
ii $f(x) = -0.01x(x - 18)^2$
iii $f(x) = -8\sin(10x)$

f **i** $f(x) = 0.09877x(x + 18)$
ii $f(x) = 0.01x(x + 18)^2$

iii $f(x) = -8\sin(-10x)$

10 a $y = 0.08x^2$

b $y = 4\cos30(x - 16) + 12$

c $y = 0.08(x - 32)^2$

d $y = 4\cos(30(x - 48)) + 12$

e $y = 0.08(x - 64)^2$

f $y = 4\cos(30(x - 80)) + 12$

g $y = 0.08(x - 96)^2$

Practice tasks (pp. 115–123)

Practice task one (pp. 115–117)

Part A

Section 1:
Local minimum at start $\Rightarrow$ takes the form
$H = ad^2(d - b)$
$(5, 5) \Rightarrow 5 = 125a - 25ab$
$(20, 20) \Rightarrow 20 = 8000a - 400ab$
$\therefore a = -0.01$ and $b = 25$
Model: $H = -0.01d^2(d - 25)$
Domain: $0 \leq d < 20$

Section 2:
Quadratic with minimum at (30, 10) $\Rightarrow$
$H = a(d - 30)^2 + 10$
$(20, 20) \Rightarrow a = 0.1$
Model: $H = 0.1(d - 30)^2 + 10$
Domain: $20 \leq d < 30$

Section 3:
Highest at (30, 10) and lowest at (36, 4) $\Rightarrow$
- Amplitude = 3 m so A = 3
- Period = 12 m $\Rightarrow B = \dfrac{360}{12} = 30$

$\therefore$ Equation takes the form
$H = 3\sin30(d - C) + D$
Model: $(30, 10) \Rightarrow H = 3\sin30(d - 27) + 7$
or $(45, 7) \Rightarrow H = 3\sin30(d - 39) + 7$
Domain: $30 \leq d < 45$

Section 4:
$(45, 7) \Rightarrow A = 42 - 6C$
$(59, 0) \Rightarrow A = -20C$
$\therefore C = -3$, and $A = 60$
Model: $H = \dfrac{60}{d - 39} - 3$
Domain: $45 \leq d \leq 59$

Part B

Parabola:
Minimum point at (59, 0) $\Rightarrow H = a(d - 59)^2$
$(45, 7) \Rightarrow a = 0.03571$
Model: $H = 0.03571(d - 59)^2$
This would give a much smoother end to the ride than the hyperbola because the gradient will

decrease as the roller coaster approaches (59, 0).

Exponential:

$(45, 7) \Rightarrow ab^{45} = 8$

$(59, 0) \Rightarrow ab^{59} = 1$

$\therefore b^{14} = 0.125$, so $b = 0.8620$ and

$H = a(0.8620)^d - 1$

$(45, 7) \Rightarrow a = 6386.5$

Model: $H = 6386.5(0.8620)^d - 1$

Like the parabolic curve, this would give a much smoother end to the ride than the hyperbola because the gradient will decrease as the roller coaster approaches (59, 0). However, it would not be as smooth as for the parabola because the horizontal asymptote for this curve is one metre below ground level.

Practice task two (pp. 118–120)

Part A

Parabola:

$(15, 0)$ and $(-15, 0) \Rightarrow$ parabola takes the form

$h = a(r - 15)(r + 15)$

$(5, 15) \Rightarrow a = -\dfrac{3}{40}$

Model for both sides: $h = -\dfrac{3}{40}(r + 15)(r - 15)$

Domain: $-15 \le r \le -5$ and $5 \le r \le 15$

Hyperbola:

RHS: centre of rotation at $(3, -3)$ and it takes the form $h = \dfrac{a}{r + b} + c \Rightarrow h = \dfrac{a}{r - 3} - 3$

$(15, 0) \Rightarrow a = 36$

Model for RHS: $h = \dfrac{36}{r - 3} - 3$

Domain: $5 \le r \le 15$

LHS: centre of rotation at $(-3, -3)$ and it takes the form $h = \dfrac{a}{r + b} + c \Rightarrow h = \dfrac{a}{r + 3} - 3$

$(-15, 0) \Rightarrow a = -36$

Model for LHS: $h = \dfrac{-36}{r - 3} - 3$

Domain: $-15 \le r \le -5$

Part B

Parabola:

Parabola will now pass through $(13, 0)$ and $(-13, 0)$ $\Rightarrow$ parabola takes the form $h = a(r - 13)(r + 13)$

$(3, 15) \Rightarrow a = -\dfrac{3}{32}$

Model for both sides: $h = -\dfrac{3}{32}(r + 13)(r - 13)$

Domain: $-13 \le r \le -3$ and $3 \le r \le 13$

Hyperbola:

RHS: Original: $h = \dfrac{36}{r - 3} - 3$

Translation of 2 units to the left $\Rightarrow h = \dfrac{36}{r - 1} - 3$

Model for RHS: $h = \dfrac{36}{r - 1} - 3$

Domain: $3 \le r \le 13$

Transformation: translation of 2 units to the left.

LHS: Original: $h = \dfrac{-36}{r + 3} - 3$

Translation of 2 units to the right $\Rightarrow h = \dfrac{-36}{r + 1} - 3$

Model for LHS: $h = \dfrac{-36}{r + 1} - 3$

Domain: $-13 \le r \le -3$

Transformation: translation of 2 units to the right.

Part C

Parabola:

Original: $h = -\dfrac{3}{40}(r + 15)(r - 15)$

Parabola now passes through

$\left(-15 + \dfrac{n}{2}, 0\right)$ and $\left(15 - \dfrac{n}{2}, 0\right)$

$\Rightarrow$ parabola takes the form

$h = a\left(r + 15 - \dfrac{n}{2}\right)\left(r - 15 + \dfrac{n}{2}\right)$

$\left(5 - \dfrac{n}{2}, 15\right) \Rightarrow a = \dfrac{15}{\left(5 - \dfrac{n}{2} - 15 + \dfrac{n}{2}\right)\left(5 - \dfrac{n}{2} + 15 - \dfrac{n}{2}\right)}$

$= \dfrac{-15}{200 - 10n}$

Translation of $\dfrac{n}{2}$ units to the left

$\Rightarrow$ model becomes:

$h = \dfrac{-15}{200 - 10n}\left(r + 15 - \dfrac{n}{2}\right)\left(r - 15 + \dfrac{n}{2}\right)$

Hyperbola:

Original: $h = \dfrac{36}{r - 3} - 3$

Translation of $\dfrac{n}{2}$ units to the left $\Rightarrow$

model becomes $h = \dfrac{36}{\left(r - 3 + \dfrac{n}{2}\right)} - 3$

Practice task three (pp. 121–123)

Part A

Carl's Concrete

Up to 5 m³

Maximum at $(5, 700) \Rightarrow C = a(v - 5)^2 + 700$

Passes through $(0, 400) \Rightarrow 400 = a(v - 0)^2 + 700$
$$\therefore a = -12$$
Model: $C = -12(v - 5)^2 + 700$
Range: $0 < v \le 5$
Over 5 and up to 10 m³
Translated up 300 and to the right by $5 \Rightarrow$
$C = -12(v - 5 - 5)^2 + (700 + 300)$
Model: $C = -12(v - 10)^2 + 1000$
Range: $5 < v \le 10$
Over 10 m³
Crosses v-axis at 0 and $40 \Rightarrow C = av(v - 40)$
Passes through $(10, 1100) \Rightarrow 1100 = a \times 10(10 - 40)$
$$\therefore a = -3.\dot{6}$$
Model: $C = -3.\dot{6}\, v(v - 40)$
Range: $10 < v \le 20$
Connie's Concrete
Up to 5 m³
$(0, 300) \Rightarrow 300 = \dfrac{a}{b} + 100$

$$\therefore a = 200b \qquad \text{①}$$

$(5, 700) \Rightarrow 700 = \dfrac{a}{5 + b} + 100$

$$\therefore a = 3000 + 600b \qquad \text{②}$$
Solving ① and ② $\Rightarrow a = -1500$ and $b = -7.5$

Model: $C = \dfrac{-1500}{v - 7.5} + 100$

Range: $0 < v \le 5$

Over 5 m³
$(5, 700) \quad \Rightarrow \qquad 700 = \dfrac{a}{5} + b$

$$\therefore a = 3500 - 5b \qquad \text{①}$$

$(10, 1100) \Rightarrow \qquad 1100 = \dfrac{a}{10} + b$

$$\therefore a = 1100 - 10b \qquad \text{②}$$
Solving ① and ② $\Rightarrow a = -4000$ and $b = 1500$

Model: $C = \dfrac{-4000}{v} + 1500$

Range: $5 < v \le 20$
Recommendations:
Less than 5 m³: Connie
Exactly 5 m³: Either
More than 5 and up to 10 m³: Carl
More than 10 m³: Connie

Part B
This question has a range of answers, so check with your teacher if yours is not like the one below.

Model takes the form $C = a \sin b(v + c) + d$
Minimum where $v = 0$ and maximum where $v = 20$

$\Rightarrow$ Period = 40
$\therefore b = 9$
Least he can afford for a load is $200, so let graph pass through $(0, 200)$.
Wants to charge less than Carl ($1466.67) or Connie ($1300) for 20 m³, so let graph pass through $(20, 1200)$.
Minimum at $(0, 200)$ and maximum at $(20, 1200) \Rightarrow$
Amplitude = 500
$\therefore a = 500$
Normal 'centre' for $y = \sin x$ is $(0, 0)$. This has moved to $(10, 700)$.
This means that the centre for $y = a \sin(bx)$ has moved 10 units to the right and 700 units up.
$\therefore c = -10$ and $d = 700$.
Model: $C = 500 \sin 9(v - 10) + 700$
Range: $0 < v \le 20$

Changing d:
With the existing values for a, b and c, the range for d is $700 \le v < 800$. However:
- If it is exactly 800 then Kauri would be charging the same amount ($1300) as Connie for 20 m³ of concrete.
- If it is less than 700, then his minimum price would be less than $200.

Changing a:
The value for a could be increased by 50 to 550. However, this would mean that his minimum charge would be reduced to $150. So to move the graph up so the minimum charge is $200, the value for d would have to increase from 700 to 750. However, this would mean that Kauri would be charging the same amount as Connie for 20 m³ of concrete.
$\therefore$ a and d could be increased by the same amount, provided it was less than 50.
Example: $C = 549 \sin 9(v - 10) + 749$ results in a minimum charge of $200 and a maximum charge for 20 m³ of $1298, which is $2 less than Connie.